AF477167

Advanced Internal Combustion Engines

Advanced Internal Combustion Engines

K. Sudhakar

and

Anil Kumar

Assistant Professors, Department of Energy,
Maulana Azad National Institute of Technology,
Bhopal – 462051, MP.

BSP **BS Publications**

A unit of **BSP Books Pvt. Ltd.**

4-4-309/316, Giriraj Lane, Sultan Bazar,
Hyderabad - 500 095 - TG
Phone : 040 - 23445605, 23445688

Published by :

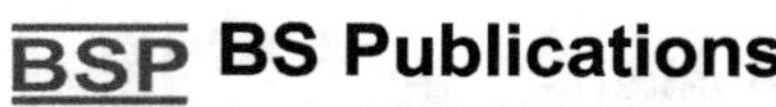 **BS Publications**

A unit of **BSP Books Pvt., Ltd.**

4-4-309/316, Giriraj Lane, Sultan Bazar,
Hyderabad - 500 095 – TG, INDIA
Phone : 040 - 23445605, 23445688
e-mail : info@bspbooks.net
www.bspbooks.net

ISBN: 978-93-85433-56-6 (HB)

Preface

The purpose of internal combustion engines is the production of mechanical power from the chemical energy contained in the fuel. The objective of writing this text book is to fill the vacuum which exists in absence of an exclusive text book on the basics and advancement in IC Engine. This book is divided in 8 chapters starting from basics of internal combustion engine to advancement and recent trends in IC Engine.

The construction and working principle of the internal combustion engine has been explained in introductory chapter.

Chapter 2 introduces the ignition and cooling system of IC Engines.

Chapter 3 explains the IC Engine Testing and Performance.

Chapter 4 introduces the combustion phenomenon in SI Engines .

Combustion in CI Engines is treated in chapter 5.

Chapter 6 deals with the pollution formation and control.

Chapter 7 deals with the conventional and alternative fuels.

Recent Trends in IC Engines are included in Chapter 8.

This book is the outcome of many years of teaching of Advanced IC Engine subject and it is intended to serve as a reference for researchers and engineers.

- The subject matter is arranged sequentially and presented in a very simple and systematic manner.
- Review questions are given in each chapter to help the students for their examinations.
- Each chapter of the book contains enough number of illustrations and figures.
- A large number of worked out examples are provided.
- Text book meets exhaustively the requirements of various B.E., B.Tech & M.E/M.Tech syllabi of various Indian universities.

We sincerely hope that this book can also be used as reference material for the teaching fraternity. We hope that the book is complete in all respects. However in any case, if it lacks certain information, authors may be contacted for constructive comments.

-Authors

Acknowledgement

Dr. K. Sudhakar

Whatever I am today is all because of my late mother Ms. K. Kalaiselvi. Her eternal guidance enlightens me in everyday of my life. She continues to provide me strength at all tough times. I bow my head with love, respect and reverence.

Dr. Anil Kumar

My wife Ms. Abhilasha, son Tijil Kumar and daughter Idika Kumar have encouraged me throughout this long and time consuming project work which took many hours away from them. Without their continuing support it would never been finish. I will always grateful to them for their patience.

Last but not the least, we express our deep gratitude to our respected parents for their blessing.

We would like to express our thanks to the various publishers and websites, who have granted permission to reproduce some figures from their sources.

We express our sincere thanks to BS Publications, Hyderabad for editing and printing this book in time.

We are highly indebted to our Director Dr. K. K. Appukuttan for his encouragement and support.

We extend our acknowledgement to Mr. Saurabh Yadav and Mr. Girish Thakur for their assistance.

-Authors

Contents

CHAPTER 1

Internal Combustion Engine

CHAPTER 2

Internal Combustion Engine System

CHAPTER 3

IC Engine Testing and Performance

CHAPTER 4

Combustion in SI Engines

CHAPTER 5

Combustion in CI Engines

CHAPTER 6

Pollutant Formation and Control

CHAPTER 7

Conventional and Alternative Fuels

CHAPTER 8

Recent Trends in IC Engines

CHAPTER 1

Internal Combustion Engine

1.1 Historical Overview of IC Engine Development

The modern reciprocating internal combustion engines have their origin in the Otto and Diesel Engines invented in the later part of 19th century. The main engine components comprising of piston, cylinder, crank-slider crankshaft, connecting road, valves and valve train, intake and exhaust system remain functionally overall similar since those in the early engines although great advancements in their design and materials have taken place during the last 100 years or so. An historical overview of IC engine development with important milestones since their first production models were built, is presented in Table 1.1.

Table 1.1 Historical Overview and Milestones in IC Engine Development

Year	Milestone
1860-1867	J. E. E. Lenoir and Nikolaus Otto developed atmospheric engine wherein combustion of fuel-air charge during first half of outward stroke of a free piston accelerating the piston which was connected to a rack assembly. The free piston would produce work during second half of the stroke creating vacuum in the cylinder and the atmospheric pressure then would push back the piston.
1876	Nikolaus Otto developed 4-stroke SI engine where in the fuel-air charge was compressed before being ignited.

Table 1.1 *Contd...*

Year	Milestone
1878	Dugald Clerk developed the first 2-stroke engine.
1882	Atkinson develops an engine having lower expansion stroke than the compression stroke for improvement in engine thermal efficiency at cost of specific engine power. The Atkinson cycle is finding application in the modern hybrid electric vehicles (HEV).
1892	Rudolf Diesel takes patent on engine having combustion by direct injection of fuel in the cylinder air heating solely by compression, the process now known as compression ignition (CI).
1896	Henry Ford develops first automobile powered by the IC engine.
1897	Rudolph Diesel developed CI engine prototype, also called as the Diesel engine.
1923	Antiknock additive tetra ethyl lead discovered by the General Motors became commercially available which provided boost to development of high compression ratio SI engines.
1957	Felix Wankel developed rotary internal combustion engine.
1981	Multipoint port fuel injection introduced on production gasoline cars.
1988	Variable valve timing and lift control introduced on gasoline cars.
1989-1990	Electronic fuel injection on heavy duty diesel introduced.
1990	Carburettor was replaced by port fuel injection on all US production cars.
1994	Direct injection stratified charge (DISC) engine powered cars came in production by Mitsubishi and Toyota.

The internal combustion engine is a heat engine that converts chemical energy in a fuel into mechanical energy, usually made available on a rotating output shaft.

Chemical energy of the fuel is first converted to thermal energy by means of combustion or oxidation with air inside the engine. This thermal energy raises the temperature and pressure of the gases within the engine and the high-pressure gas then expands against the mechanical mechanisms of the engine. This expansion is converted by the mechanical linkages of the engine to a rotating crankshaft, which is the output of the engine. Schematic view of IC Engine is shown in Fig. 1.1.

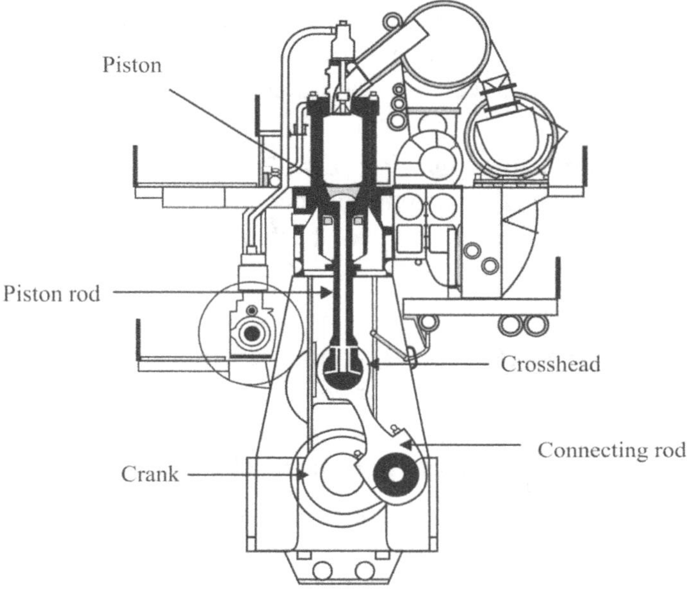

Fig. 1.1 Schematic View of IC Engine

The crankshaft, in turn, is connected to a transmission and/or power train to transmit the rotating mechanical energy to the desired final use. For engines this will often be the propulsion of a vehicle (i.e., automobile, truck, locomotive, marine vessel, or airplane). Other applications include stationary engines to drive generators or pumps, and portable engines for things like chain saws and lawn mowers.

Most internal combustion engines are reciprocating engines having pistons that reciprocate back and forth in cylinders internally within the engine.

1.2   Classification of Internal Combustion Engine

Internal combustion engines can be classified in a number of different ways:

(i) Types of Ignition

    (a) *Spark Ignition (SI):* An SI engine starts the combustion process in each cycle by use of a spark plug. The spark plug gives a high-voltage electrical discharge between two electrodes which ignites the air-fuel mixture in the combustion chamber surrounding the plug. In early engine development, before the invention of the electric spark plug, many forms of torch holes were used to initiate combustion from an external flame.

(b) *Compression Ignition (CI):* The combustion process in a CI engine starts when the air-fuel mixture self-ignites due to high temperature in the combustion chamber caused by high compression.

(ii) Engine Cycle

(a) *Four-Stroke Cycle:* A four-stroke cycle experiences four piston movements over two engine revolutions for each cycle.

(b) *Two-Stroke Cycle:* A two-stroke cycle has two piston movements over one engine revolution for each cycle.

Three-stroke cycles and six-stroke cycles were also tried in early engine development.

(iii) Valve Location

(a) Valves in head (overhead valve), also called I Head engine.

(b) Valves in block (flat head), also called L Head engine. Some historic engines with valves in block had the intake valve on one side of the cylinder and the exhaust valve on the other side. These were called T Head engines.

(c) One valve in head (usually intake) and one in block, also called F Head engine; this is much less common.

(iv) Basic Design

(a) *Reciprocating:* Engine has one or more cylinders in which pistons reciprocate back and forth. The combustion chamber is located in the closed end of each cylinder. Power is delivered to a rotating output crankshaft by mechanical linkage with the pistons.

(b) *Rotary:* Engine is made of a block (stator) built around a large non-concentric rotor and crankshaft. The combustion chambers are built into the non-rotating block.

(v) Position and Number of Cylinders of Reciprocating Engines

(a) *Single Cylinder:* Engine has one cylinder and piston connected to the crankshaft.

(b) *In-Line:* Cylinders are positioned in a straight line, one behind the other along the length of the crankshaft. They can consist of 2 to 11 cylinders or possibly more. In-line four-cylinder engines are very common for automobile and other applications. In-line six and eight cylinders are historically common automobile engines. In-line engines are sometimes called straight (e.g., straight six or straight eight).

(c) *V Engine:* Two banks of cylinders at an angle with each other along a single crankshaft. The angle between the banks of cylinders can be anywhere from 15° to 120°, with 60°-90° being common. V engines have even numbers of cylinders from 2 to 20 or more. V6s and V8s are common automobile engines, with V12s and V16s (historic) found in some luxury and high-performance vehicles.

(d) *Opposed Cylinder Engine:* Two banks of cylinders opposite each other on a single crankshaft (a V engine with a 180°V). These are common on small aircraft and some automobiles with an even number of cylinders from two to eight or more. These engines are often called flat engines (e.g., flat four).

(e) *W Engine:* Same as a V engine except with three banks of cylinders on the same crankshaft. Not common, but some have been developed for racing automobiles, both modern and historic. Usually 12 cylinders with about a 60° angle between each bank.

(f) *Opposed Piston Engine:* Two pistons in each cylinder with the combustion chamber in the center between the pistons. A single-combustion process causes two power strokes at the same time, with each piston being pushed away from the center and delivering power to a separate crankshaft at each end of the cylinder. Engine output is either on two rotating crankshafts or on one crankshaft incorporating complex mechanical linkage.

(g) *Radial Engine:* Engine with pistons positioned in a circular plane around the central crankshaft. The connecting rods of the pistons are connected to a master rod which, in turn, is connected to the crankshaft. A bank of cylinders on a radial engine always has an odd number of cylinders ranging from 3 to 13 or more. Operating on a four-stroke cycle, every other cylinder fires and has a power stroke as the crankshaft rotates, giving a smooth operation. Many medium and large-size propeller-driven aircraft use radial engines. For large aircraft, two or more banks of cylinders are mounted together, one behind the other on a single crankshaft, making one powerful, smooth engine. Very large ship engines exist with up to 54 cylinders, six banks of 9 cylinders each.

(vi)   Air Intake Process

    (a) *Naturally Aspirated:* No intake air pressure boost system exists.

    (b) *Supercharged:* Intake air pressure increased with the compressor driven off of the engine crankshaft.

    (c) *Turbocharged:* Intake air pressure increased with the turbine-compressor driven by the engine exhaust gases.

    (d) *Crankcase Compressed:* Two-stroke cycle engine uses the crankcase as the intake air compressor. Limited development work has also been done on design and construction of four-stroke cycle engines with crankcase compression.

(vii)   Method of Fuel Input for SI Engines

    (a)   Carbureted.

    (b)   *Multipoint Port Fuel Injection:* One or more injectors at each cylinder intake.

    (c)   *Throttle Body Fuel Injection:* Injectors upstream in intake manifold.

(viii) Fuel Used

    (a)   Gasoline.

    (b)   Diesel Oil or Fuel Oil.

    (c)   Gas, Natural Gas, Methane.

    (d)   LPG.

    (e)   Alcohol-Ethyl, Methyl.

    (f)   *Dual Fuel:* There are a number of engines that use a combination of two or more fuels. Some, usually large, CI engines use a combination of methane and diesel fuel. These are attractive in third world developing countries because of the high cost of diesel fuel. The engine which runs on Combined gasoline-alcohol fuels are called dual fuel engines.

(g) *Gasohol:* Common fuel is consisting of 90% gasoline and 10% alcohol.

(ix)   **Application**

(a) Automobile, Truck, Bus.

(b) Locomotive.

(c) Stationary.

(d) Marine.

(e) Aircraft.

(f) Small Portable, Chain Saw, Model Airplane.

(x)   **Type of Cooling**

(a) Air Cooled.

(b) Liquid Cooled, Water Cooled.

1.3   Nomenclature

The following terms are commonly used in IC engine technology literature and also shown in Fig. 1.2.

Spark Ignition (SI): An engine in which the combustion process in each cycle is started by use of a spark plug.

Compression Ignition (CI): An engine in which the combustion process starts when the air-fuel mixture self-ignites due to high temperature in the combustion chamber caused by high compression. CI engines are often called Diesel engines.

Top-Dead-Center (TDC): Position of the piston when it stops at the furthest point away from the crankshaft. Top, because this position is at the top of most engines (not always), and dead because the piston stops at this point. Because in some engines top-dead-center is not at the top of the engine (e.g., horizontally opposed engines, radial engines, etc.,) some sources call this position Head-End-Dead-Center (HEDC).

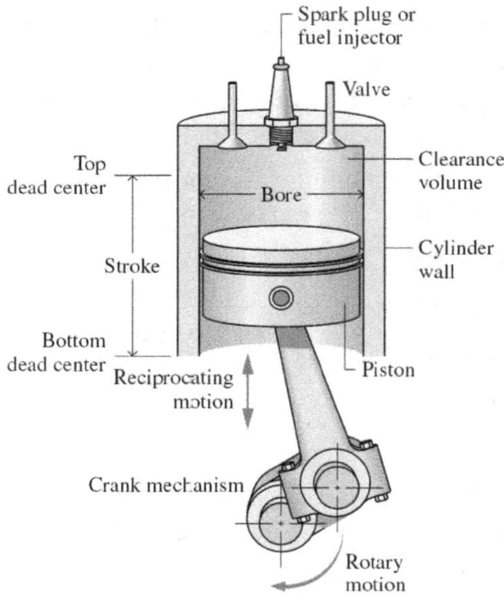

Fig. 1.2 Term Relating in IC Engine

Bottom-Dead-Center (BDC): Position of the piston when it stops at the point closest to the crankshaft. Some sources call this Crank-End-Dead-Center (CEDC) because it is not always at the bottom of the engine.

Bore (D): Diameter of the cylinder or diameter of the piston face, which is the same minus a very small clearance.

Stroke (L): Movement distance of the piston from one extreme position to the other i.e., TDC to BDC or BDC to TDC. It is equal to twice the radius of crank circle.

Clearance Volume: Minimum volume in the combustion chamber with piston at TDC.

Displacement or Displacement Volume: Volume displaced by the piston as it travels through one stroke. Displacement can be given for one cylinder or for the entire engine (one cylinder times number of cylinders). Some literature calls this *Swept Volume.*

$$V_s = \frac{\pi D^2 L}{4}$$

Compression Ratio (CR): It is the ratio of total volume to clearance volume.

1.4 Parts of Internal Combustion Engine

The following are the major components found in most reciprocating internal combustion engines (Fig. 1.3).

- **(i)** **Block:** Body of engine containing the cylinders, made of cast iron or aluminum. The block of water-cooled engines includes a water jacket cast around the cylinders. On air-cooled engines, the exterior surface of the block has cooling fins.

- **(ii)** **Camshaft:** Rotating shaft used to push open valves at the proper time in the engine cycle, either directly or through mechanical or hydraulic linkage (push rods, rocker arms, tappets). Most modern automobile engines have one or more camshafts mounted in the engine head (overhead cam). Camshafts are generally made of forged steel or cast iron and are driven off the crankshaft by means of a belt or chain (timing chain).

- **(iii)** **Carburetor:** Venturi flow device which meters the proper amount of fuel into the air flow by means of a pressure differential.

- **(iv)** **Catalytic Converter:** Chamber mounted in exhaust flow containing catalytic material that promotes reduction of emissions by chemical reaction.

- **(v)** **Combustion Chamber:** The end of the cylinder between the head and the piston face where combustion occurs. The size of the combustion chamber continuously changes from a minimum volume when the piston is at TDC to a maximum when the piston is at BDC.

- **(vi)** **Connecting Rod:** Rod connecting the piston with the rotating crankshaft, usually made of steel or alloy forging in most engines but may be aluminum in some small engines.

- **(vii)** **Connecting Rod Bearing:** Bearing where connecting rod fastens to crankshaft.

- **(viii)** **Cooling Fins:** Metal fins on the outside surfaces of cylinders and head of an air-cooled engine. These extended surfaces cool the cylinders by conduction and convection.

- **(ix)** **Crankcase:** Part of the engine block surrounding the rotating crankshaft. In many engines, the oil pan makes up part of the crankcase housing.

- **(x)** **Crankshaft:** Rotating shaft through which engine work output is supplied to external systems. The crankshaft is connected to the engine block with the main bearings. It is rotated by the reciprocating pistons through connecting rods connected to the crankshaft. Most crankshafts are made of forged steel, while some are made of cast iron.

(xi)   **Cylinders:** The circular cylinders in the engine block in which the pistons reciprocate back and forth. The walls of the cylinder have highly polished hard surfaces. Cylinders may be machined directly in the engine block, or a hard metal (drawn steel) sleeve may be pressed into the softer metal block.

(xii)   **Exhaust Manifold:** Piping system which carries exhaust gases away from the engine cylinders, usually made of cast iron.

(xiii)   **Exhaust System:** Flow system for removing exhaust gases from the cylinders, treating them, and exhausting them to the surroundings. It consists of exhaust manifold which carries the exhaust gases away from the engine, a thermal or catalytic converter to reduce emissions, a muffler to reduce engine noise.

(xiv)   **Flywheel:** Rotating mass with a large moment of inertia connected to the crank-shaft of the engine. The purpose of the flywheel is to store energy and furnish a large angular momentum that keeps the engine rotating between power strokes and smoothes out engine operation.

(xv)   **Fuel Injector:** A pressurized nozzle that sprays fuel into the incoming air on SI engines or into the cylinder on CI engines. On SI engines, fuel injectors are located at the intake valve ports on multipoint port injector systems and upstream at the intake manifold inlet on throttle body injector systems. In a few SI engines, injectors spray directly into the combustion chamber.

(xvi)   **Fuel Pump:** Electrically or mechanically driven pump to supply fuel from the fuel tank (reservoir) to the engine. Many modern automobiles have an electric fuel pump mounted submerged in the fuel tank. Some small engines and early automobiles had no fuel pump, relying on gravity feed.

(xvii)   **Head:** The piece which closes the end of the cylinders, usually containing part of the clearance volume of the combustion chamber. The head is usually made of cast iron or aluminum, and bolts to the engine block. In some engines, the head is one piece with the block. The head contains the spark plugs in SI engines and the fuel injectors in CI engines. Most modern engines have the valves in the head.

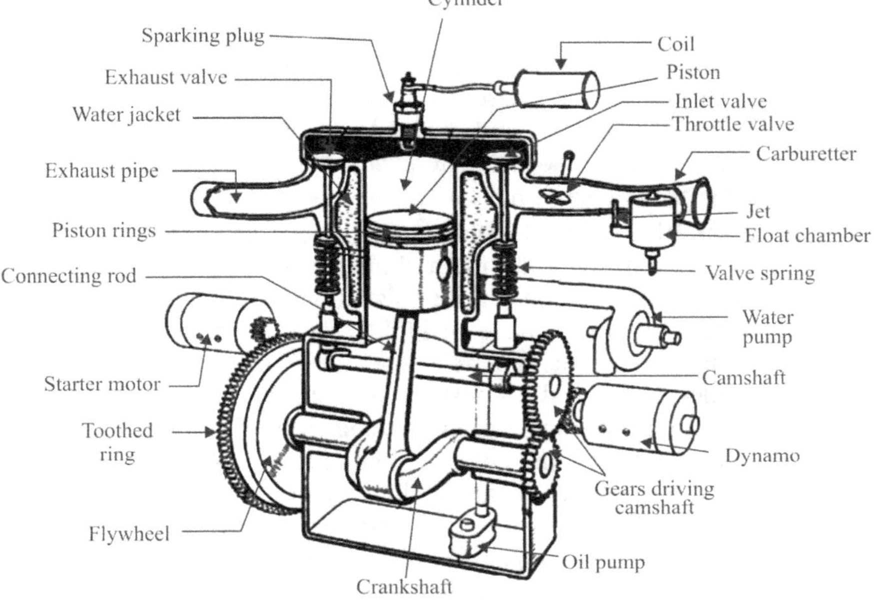

Fig. 1.3 Construction of Four-Stroke Cycle SI Engine

(xviii) Head Gasket: Gasket which serves as a sealant between the engine block and head where they bolt together. They are usually made in sandwich construction of metal and composite materials.

(xix) Intake Manifold: Piping system which delivers incoming air to the cylinders usually made of cast metal, plastic, or composite material. In most SI engines, fuel is added to the air in the intake manifold system either by fuel injectors or with a carburetor.

(xx) Oil Pump: Pump used to distribute oil from the oil sump to required lubrication points. The oil pump can be electrically driven, but is most commonly mechanically driven by the engine. Some small engines do not have an oil pump and are lubricated by splash distribution.

(xxi) Oil Dump: Reservoir for the oil system of the engine, commonly part of the crankcase.

(xxii) Piston: The cylindrical-shaped mass that reciprocates back and forth in the cylinder, transmitting the pressure forces in the combustion chamber to the rotating crankshaft. The top of the piston is called the crown and the sides are called the skirt. The face on the crown makes up one wall of the combustion chamber and may be a flat or highly contoured surface. Pistons are made of cast iron, steel, or aluminum. Iron and steel pistons can have sharper corners because of their higher strength. Aluminum pistons are lighter and have less mass inertia. Sometimes synthetic or composite materials are used for the body of the piston. Some pistons have a ceramic coating on the face.

(xxiii) **Piston Rings:** Metal rings that fit into circumferential grooves around the piston and form a sliding surface against the cylinder walls. Near the top of the piston are usually two or more compression rings made of highly polished hard chrome steel. The purpose of these is to form a seal between the piston and cylinder walls and to restrict the high-pressure gases in the combustion chamber from leaking past the piston into the crankcase. Below the compression rings on the piston is at least one oil ring, which assists in lubricating the cylinder walls and scrapes away excess oil to reduce oil consumption.

(xxiv) **Push Rods:** Mechanical linkage between the camshaft and valves on overhead valve engines with the camshaft in the crankcase.

(xxv) **Spark Plug:** Electrical device used to initiate combustion in an SI engine by creating a high-voltage discharge across an electrode gap. Spark plugs are usually made of metal surrounded with ceramic insulation.

(xxvi) **Super Charger:** Mechanical compressor powered off of the crankshaft, used to compress incoming air of the engine.

(xxvii) **Throttle:** Butterfly valve mounted at the upstream end of the intake system, used to control the amount of air flow into an SI engine. Some small engines and stationary constant-speed engines have no throttle.

(xxviii) **Turbo Charger:** Turbine-compressor used to compress incoming air into the engine.

The turbine is powered by the exhaust flow of the engine and thus takes very little useful work from the engine.

(xxix) **Valves:** Used to allow flow into and out of the cylinder at the proper time in the cycle. Most engines use poppet valves, which are spring loaded closed and pushed open by camshaft action. Valves are mostly made of forged steel. Many two-stroke cycle engines have ports (slots) in the side of the cylinder walls instead of mechanical valves.

(xxx) Wrist Pin: Pin fastening the connecting rod to the piston (also called the piston pin).

1.5 Spark Ignition Engine (Petrol Engine)

1.5.1 Working of 2-Stroke SI Engine

(i) **Combustion:** With the piston at TDC combustion occurs very quickly, raising the temperature and pressure to peak values, almost at constant volume (Fig. 1.4(e)).

(ii) **First Stroke (Expansion Stroke or Power Stroke):** Very high pressure created by the combustion process forces the piston down in the power stroke. The expanding volume of the combustion chamber causes pressure and temperature to decrease as the piston travels towards BDC (Fig. 1.4 (a)).

(iii) **Exhaust Blowdown:** At about BDC, the exhaust valve opens and blowdown occurs. The exhaust valve may be a poppet valve in the cylinder head, or it may be a slot in the side of the cylinder which is uncovered as the piston approaches BDC. After blowdown the cylinder remains filled with exhaust gas at lower pressure (Fig. 1.4(b)).

(iv) **Intake and Scavenging:** When blowdown is nearly complete, at about BDC, the intake slot on the side of the cylinder is uncovered and intake air-fuel enters under pressure. Fuel is added to the air with either a carburetor or fuel injector. This incoming mixture pushes much of the remaining exhaust gases out the open exhaust valve and fills the cylinder with a combustible air-fuel mixture, a process called scavenging. The piston passes BDC and very quickly covers the intake port and then the exhaust port (or the exhaust valve closes). The higher pressure at which the air enters the cylinder is established in one of two ways. Large two-stroke cycle engines generally have a supercharger, while small engines will intake the air through the crankcase. On these engines the crankcase is designed to serve as a compressor in addition to serving its normal function (Fig. 1.4(c)).

(v) **Second Stroke (Compression Stroke):** With all valves (or ports) closed, the piston travels towards TDC and compresses the air-fuel mixture to a higher pressure and temperature. Near the end of the compression stroke, the spark plug is fired; by the time the piston gets to IDC, combustion occurs and the next engine cycle begins (Fig. 1.4(d)).

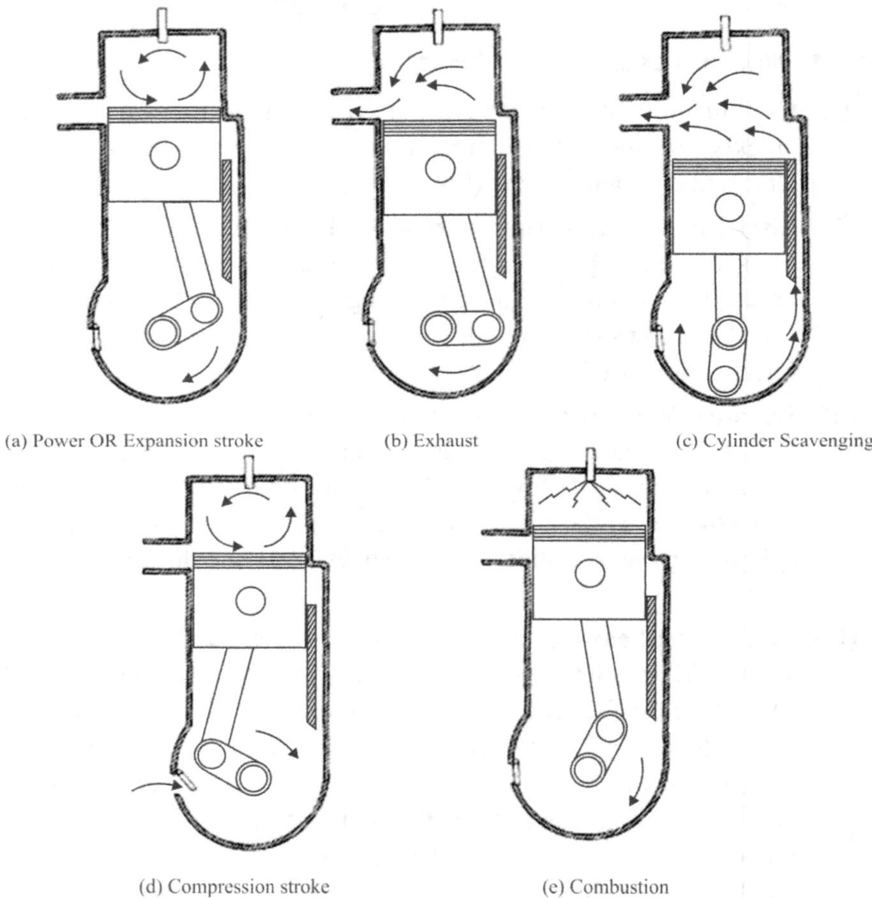

(a) Power OR Expansion stroke   (b) Exhaust   (c) Cylinder Scavenging

(d) Compression stroke   (e) Combustion

Fig. 1.4 2-Stroke SI Engine Operating Cycle

The theoretical indicator diagram for a two stroke cycle pertrol engines is shown in Fig. 1.5

Process 1-2: isentropic compression of the charge in the cylinder. The charge inside the cylinder is compressed and the piston moves from bottom dead center to top dead center. This comprises first stroke of the engine.

Process 2-3: combustion at constant volume. This process takes place as process of constant volume and increase in pressure. In this process spark is ignited by spark plug inside the cylinder and fuel is burnt.

Process 3-4: isentropic expansion. The burnt fuel exerts pressure and moves the piston to bottom dead center. The gas expands in this process. This comprises the second stroke and the power stroke of the engine.

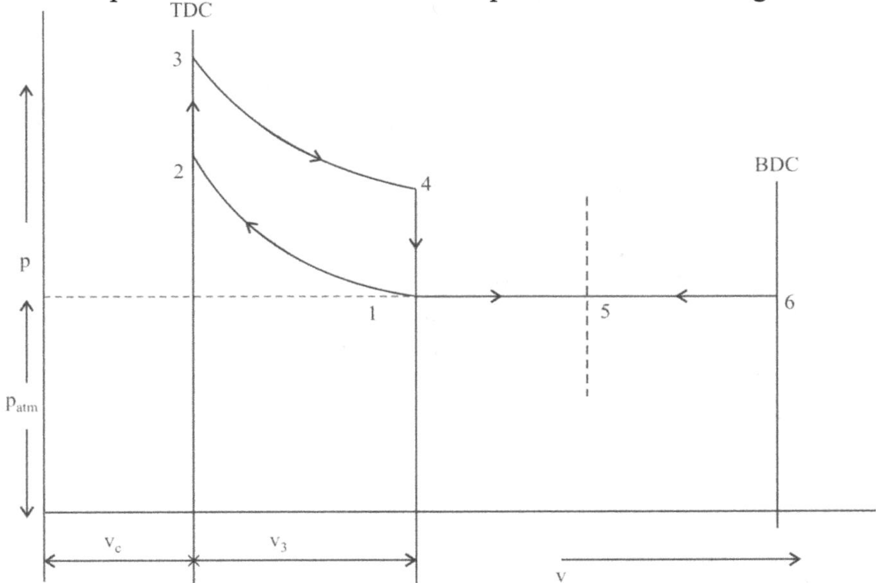

Fig. 1.5 Theoretical p-V Diagram of a Two Stroke Otto Cycle Engine

Process 4-1: Sudden release of burnt gases to the atmosphere as the exhaust port opens. This process takes place as process of constant volume and decrease in pressure. In this process, the burnt gas is exhausted outside the cylinder.

Process 1-6& 6-1: Sweeping out the exhaust gases to the atmosphere at the atmospheric pressure and charging the cylinder with charge.

Process 5-6 & 6-5: Charging the cylinder with charge through   the transfer ports and scavenging   through exhaust ports.

The point "5" indicates the opening of the inlet or transfer ports.

Also the compression of the charge starts   from point "1" instead of point "6" Therefore, the effective stroke (compression stroke) is less than the actual stroke.

The actual indicator diagram for a two stroke cycle petrol engine is shown in Fig. 1.6. The section is shown by the line 1-2-3 (from the instant transfer port opens (TPO) and transfer port closes). During the section stage the exhaust port is also open. In the first half of suction stage, the volume of fuel-air mixture and burnt gases increases. This happens as the piston moves from 1 to 2. In the second half of suction stage, the volume of charge and burnt gases decreases. This happens as the piston moves upwards from 2 to 3. A little beyond 3, the exhaust port closes (EPC) at 4. Now the charge inside the engine cylinder is compressed which is shown by the line 4-5. At the end of compression, there is an increase in pressure inside the engine cylinder. Shortly before the end of compression, the charge is ignited with the help of spark plug as shown in Fig. 14 e. The sparking suddenly increases temperature and pressure of the product of combustion. But the volume, practically, remains constant as shown by the line 5-6. The expansion is shown by the line 6-7. Now the exhaust port opens at 7, and the burnt gases are exhausted into the atmosphere through the exhaust port. It reduces pressure. As the piston moving towards BDC, therefore volume of burnt gases increases from 7 to 1. At 1 the transfer port opens (TPN) and suction starts.

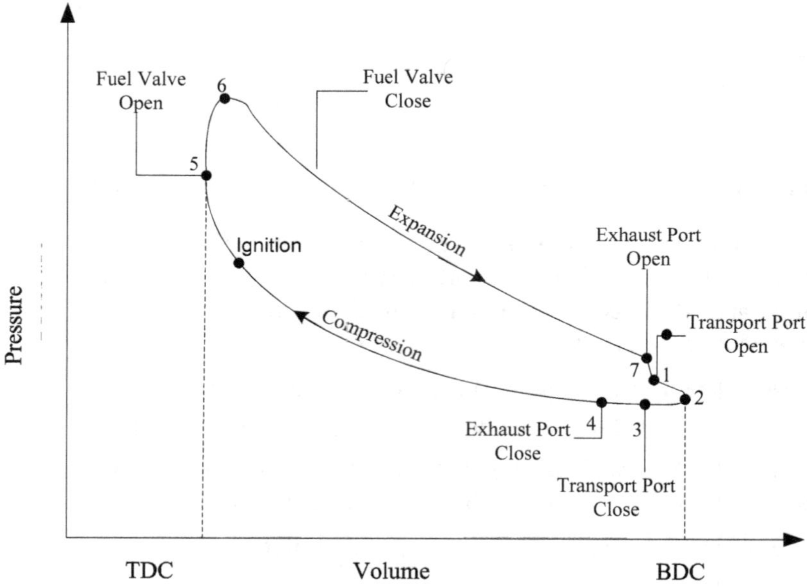

Fig. 1.6 Actual P-V Diagram of a Two Stroke Otto Cycle Engine

1.5.2 Working of 4-Stroke SI Engine

(i) **First Stroke (Intake Stroke or Induction):** The piston travels from TDC to BDC with the intake valve open and exhaust valve closed. This creates an increasing volume in the combustion chamber, which in turn creates a vacuum. The resulting pressure differential causes air to be pushed into the cylinder. As the air passes through the intake system, fuel is added to it in the desired amount by means of fuel injectors or a carburetor (Fig. 1.7(a)).

(ii) Second Stroke: Compression Stroke When the piston reaches BDC, the intake valve closes and the piston travels back to TDC with all valves closed. This compresses the air-fuel mixture, raising both the pressure and temperature in the cylinder. The finite time required to close the intake valve. Near the end of the compression stroke, the spark plug is fired and combustion is initiated (Fig. 1.7(b)).

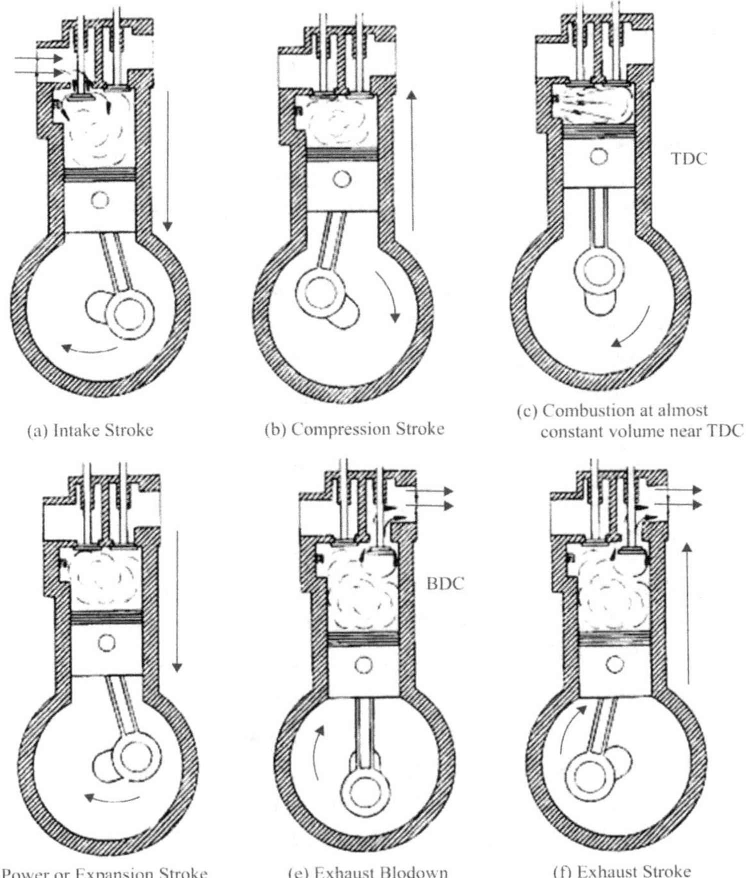

Fig. 1.7 4-Stroke SI Engine Operating Cycle

(iii) Combustion: Combustion of the air-fuel mixture occurs in a very short but finite length of time with the piston near TDC (i.e., nearly constant-volume combustion). It starts near the end of the compression stroke. Combustion changes the composition of the gas mixture to that of exhaust products and increases the temperature in the cylinder to a very high peak value. This, in turn, raises the pressure in the cylinder to a very high peak value (Fig. 1.7(c)).

(iv) Third Stroke (Expansion Stroke or Power Stroke): With all valves closed, the high pressure created by the combustion process pushes the piston away from TDC. This is the stroke which produces the work output of the engine cycle. As the piston travels from TDC to BDC, cylinder volume is increased, causing pressure and temperature to drop (Fig. 1.7(d)).

(v) Exhaust Blowdown: Late in the power stroke, the exhaust valve is opened and exhaust blowdown occurs. Pressure and temperature in the cylinder are still high relative to the surroundings at this point, and a pressure differential is created through the exhaust system which is open to atmospheric pressure. This pressure differential causes much of the hot exhaust gas to be pushed out of the cylinder and through the exhaust system when the piston is near BDC. This exhaust gas carries away a high amount of enthalpy, which lowers the cycle thermal efficiency. Opening the exhaust valve before BDC reduces the work obtained during the power stroke but is required because of the finite time needed for exhaust blowdown (Fig. 1.7(e)).

(vi) Fourth Stroke (Exhaust Stroke): By the time the piston reaches BDC, exhaust blowdown is complete, but the cylinder is still full of exhaust gases at approximately atmospheric pressure. With the exhaust valve remaining open, the piston now travels from BDC to TDC in the exhaust stroke. This pushes most of the remaining exhaust gases out of the cylinder into the exhaust system at about atmospheric pressure, leaving only that trapped in the clearance volume when the piston reaches TDC. Near the end of the exhaust stroke, the intake valve starts to open, so that it is fully open by TDC when the new intake stroke starts the next cycle. Near TDC the exhaust valve starts to close (Fig. 1.7(f)).

Theoretical indicator diagram for a four-stroke cycle petrol engine: In the above operation, the following assumption were made-

(i)   Suction and exhaust take place at atmospheric pressure.

(ii)   Suction and exhaust take place at $180°$ rotation of crank.

(iii)   Compression and expansion also take place at $180°$ rotation of crank.

(iv)   Compression and expansion are isentropic.

(v)   The combustion takes place instantaneously at constant volume at the end of compression stroke.

(vi)   Pressure suddenly falls to the atmospheric pressure at end of expansion stroke.

With these assumptions the working of 4-Stroke Otto cycle engine on p-V diagram as shown in Fig. 1.8.

Process 5-1 (Suction Stroke): In this process fresh air and fuel mixture i.e., charge is passed inside the cylinder. The piston moves from top dead center to bottom dead center. This comprises the first stroke of the engine.

Process 1-2 (Compression Stroke): In this process the charge is compressed and piston is moved to top dead center. This comprises the second stroke of the engine.

Process 2-3 (Instantaneous-Combustion): In this process spark plug ignites the spark and the fuel is burnt. This process is of constant volume and increase in pressure.

Process 3-4 (Expansion Stroke): In this process the burnt fuel expands itself and exerts pressure on the piston. The piston moves from top dead center to bottom dead center. This comprises the third stroke of the engine and the power stroke.

Process 4-1(Sudden Fall in Pressure): In this process the burnt gas is exhausted out and the pressure decreases with constant volume.

Process 1-5 (Exhaust Stroke): In this process the burnt gas is completely moved out of the cylinder by the action of piston. Piston moves from bottom dead center to top dead center. This comprises the fourth stroke of the engine.

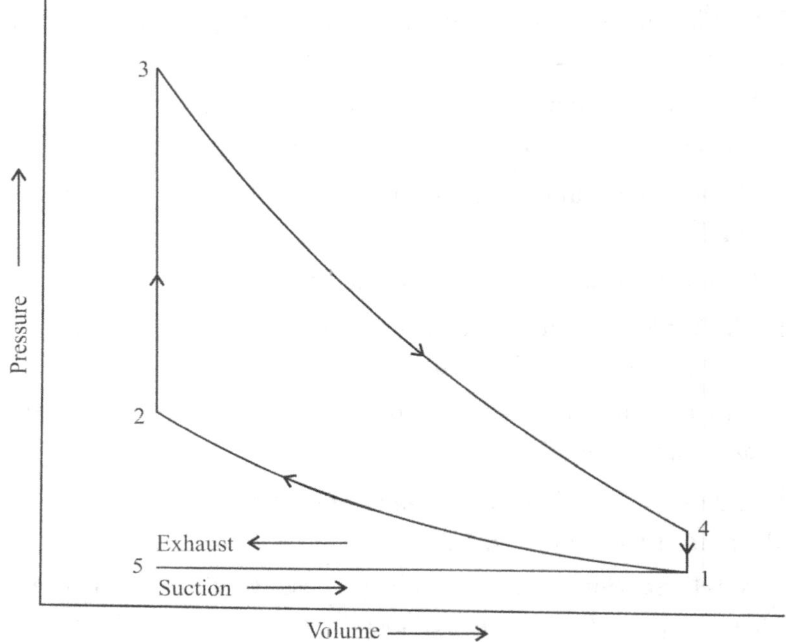

Fig. 1.8 Theoretical p-V Diagram of a Four-Stroke Otto Cycle Engine

Actual indicator (p-V) diagram for a four stroke cycle petrol cycle: In the above operations, all the ideal conditions are assumed but in practice, the actual conditions differ from the ideal as described below (Fig. 1.9).

(i) The suction of mixture in the cylinder is possible only if the pressure inside the cylinder is below atmospheric pressure.

(ii) The burnt gases can be pushed out into the atmosphere only if the pressure of the exhaust gases is above atmospheric pressure.

(iii) The combustion and expansion do not follow the isentropic law, as there will be heat exchange during process.

(iv) Sudden pressure rise is not possible after ignition as combustion take sometime for completion and actual pressure rise is less than theoretical considered. The pressure increase takes place through some crank rotation, or increase in volume.

(v) Sudden pressure release after the opening of expansion valve is not possible and also takes place through some crank rotation.

It may be noted that line 5-1 is below the atmospheric pressure line. This is due to the fact that owing to restricted area of the inlet passage the entering fuel mixture cannot cope with the speed of the piston. The exhaust line 4-5 is slightly above the atmospheric pressure line. This is due to the fact that owing to restricted area of the exhaust passage, which does not allow exhaust gases to leave the engine cylinder quickly.

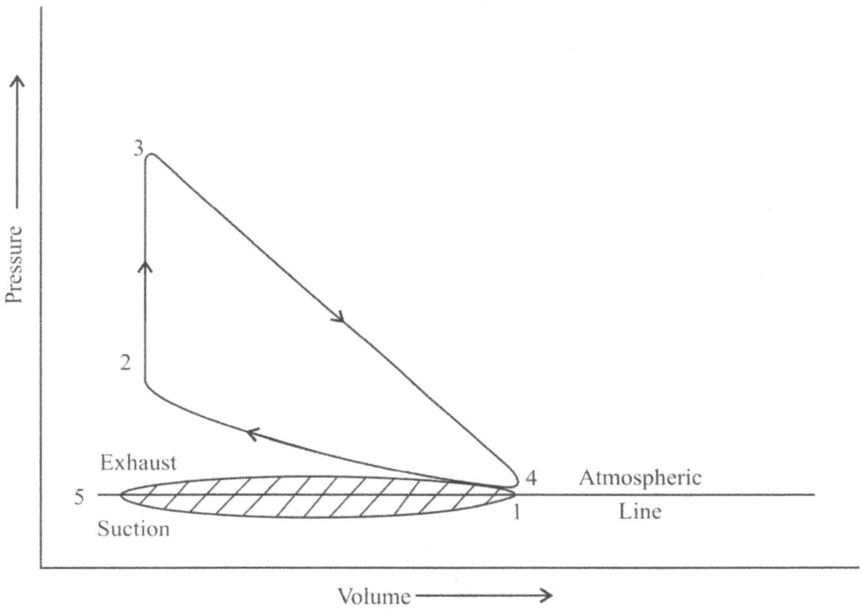

Fig. 1.9 Actual p-V Diagram of a Four-Stroke Otto Cycle Engine

The loop, which has area 4-5-1 is, called negative loop. It gives the pumping loss due to admission of fuel air mixture and removal of exhaust gases. The area 1-2-3-4 is the gross work (Fig. 1.9).

1.6 Compression Ignition Engine (Diesel Engine)

1.6.1 Working of 2-Stroke CI Engine

A two-stroke diesel engine (Fig. 1.10) shares the same operating principles as other internal combustion engines. It has all of the advantages that other diesel engines have over gasoline engines. A two-stroke diesel engine does not produce as much power as a four-stroke diesel engine; however, it runs smoother than the four-stroke diesel. This is because it generates a power stroke each time the piston moves downward; that is, once for each crankshaft revolution. The two-stroke diesel engine has a less complicated valve train because it does not use intake valves. Instead, it requires a supercharger to force air into the cylinder and to force exhaust gases out, because the piston cannot do this naturally as in four-stroke engines. The two-stroke diesel takes in air and discharges exhaust through a system called scavenging. Scavenging begins with the piston at bottom dead center. At this point, the intake ports are uncovered in the cylinder wall and the exhaust valve is open. The supercharger forces air into the cylinder, and, as the air is forced in, the burned gases from the previous operating cycle are forced out (Fig. 1.11).

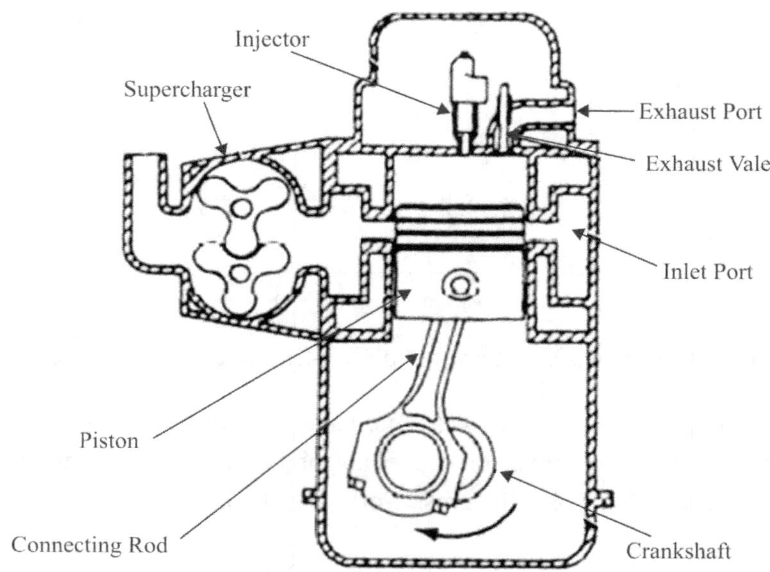

Fig. 1.10 Two-Stroke Cycle Diesel Engine

Compression Stroke: As the piston moves towards top dead center, it covers the intake ports. The exhaust valves close at this point and seals the upper cylinder. As the piston continues upward, the air in the cylinder is tightly compressed (Fig. 1.11). As in the four-stroke cycle diesel, a tremendous amount of heat is generated by the compression.

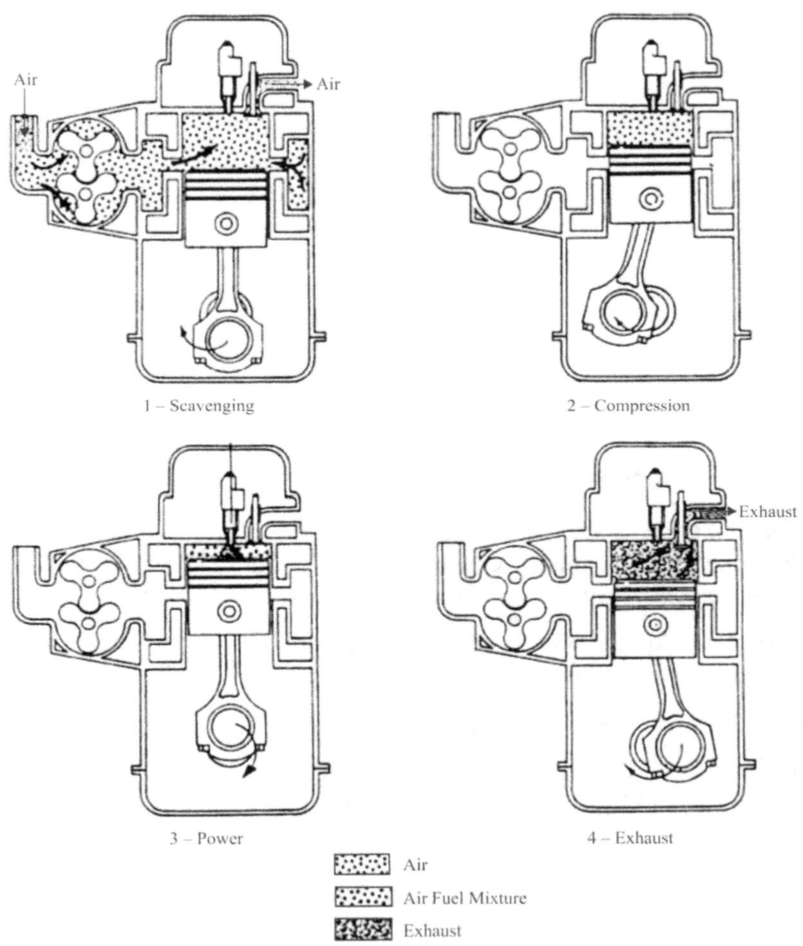

Fig. 1.11 Two-Stroke CI Engine Operating Cycle

Power Stroke: As the piston reaches top dead center, the compression stroke ends. Fuel is injected at this point and the intense heat of the compression causes the fuel to ignite. The burning fuel pushes the piston down, giving power to the crankshaft. The power stroke ends when the piston gets down to the point where the intake ports are uncovered. At about this point, the exhaust valve opens and scavenging begins again, as shown in Fig. 1.11.

Fig. 1.12 shows theoretical p-V diagram of two-stroke diesel cycle engine.

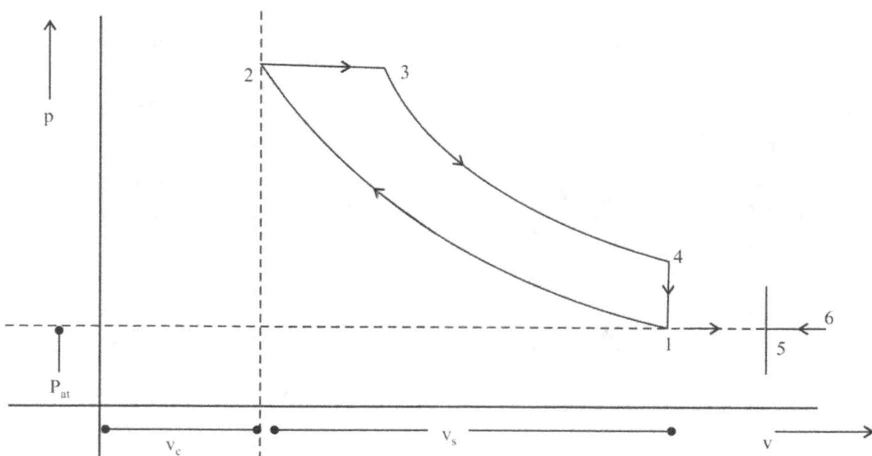

Fig. 1.12 Theoretical p-V Diagram of a Two Stroke Diesel Cycle Engine

Process 1-2: Isentropic compression of the air in the cylinder. The air inside the cylinder is compressed and the piston moves from bottom dead center to top dead center. This comprises the first stroke of the engine.

Process 2-3: Combustion at constant Pressure. In this process diesel is pumped inside the cylinder by the help of fuel pump and fuel is burnt.

Process 3-4: Isentropic Expansion of the burnt fuel in the cylinder. The burnt fuel exerts pressure and moves the piston to bottom dead center. The gas expands in this process. This comprises the second stroke and the power stroke of the engine.

Process 4-1: Sudden release of burnt gases to the atmosphere as the exhaust port opens. This process takes place as process of constant volume and decrease in pressure. In this process the burnt gas is exhausted outside the cylinder.

Process 1-6 & 6-1: Sweeping out the exhaust gases to the atmosphere.

Process 5-6 & 6-5: Charging the cylinder with fresh charge through the transfer ports and scavenging action takes place.

Fig. 1.13 shows the actual indicator diagram for a two stroke cycle diesel engine. The suction is shown by the line 1-2-3, from the instant transfer ports opens and transfer port closes. During the suction stage, the exhaust port is opens. In the first half of suction stage, the volume of air and burnt gases increases. This happens as the piston moves from 1-2. In the second half of the suction stage, the volume of air and burnt gases decreases. This happens as the piston moves upwards from 2-3. A little beyond 3, the exhaust port closes at 4. Now the air inside the engine cylinder is compressed which is shown by the line 4-5. At the end of compression there is an increases in pressure inside the engine cylinder, shortly before the end of compression, the fuel valve opens and the fuel is injected into the engine cylinder. The fuel is ignited by high temperature of the compressed air. The ignition suddenly increases volume and temperature of the products of combustion. But the pressure, practically, remains constant as shown by the line 5-6. The expansion is shown by the line 6-7. Now the exhaust port opens at 7, and the burnt gases are exhausted into the atmosphere through the exhaust port. It reduces pressure. As the piston moving towards BDC, therefore volume of burnt gases increases from 7 to 1. At 1 the transfer port opens (TPO) and suction starts.

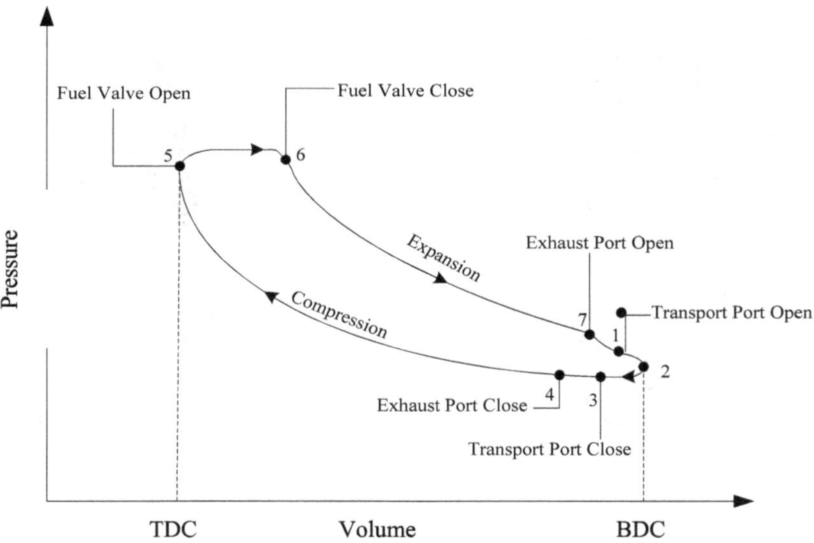

Fig. 1.13 Actual P-V Diagram of a Two Stroke Diesel Cycle Engine

1.6.2 Working of 4-Stroke Diesel Engine

Intake Stroke: The piston is at top dead center at the beginning of the intake stroke, and, as the piston moves downward, the intake valve opens. The downward movement of the piston draws air into the cylinder, and, as the piston reaches bottom dead center, the intake valve closes (Fig. 1.14(a)).

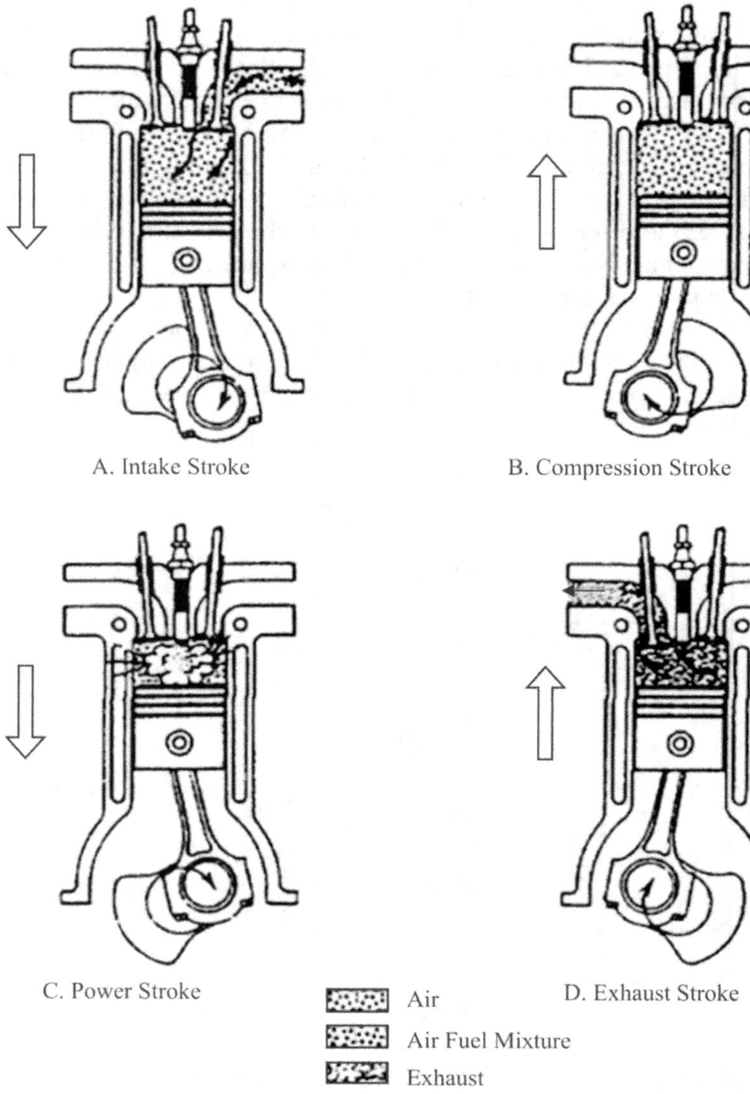

Fig. 1.14 Four Stroke CI Engine Operating Cycle

Compression Stroke: The piston is at bottom dead center at the beginning of the compression stroke, and, as the piston moves upward, the air compresses. As the piston reaches top dead center, the compression stroke ends (Fig. 1.14(b)).

Power Stroke: The piston begins the power stroke at top dead center. The air is compressed to as much as 500 psi and at a compressed temperature of approximately 1000°F. At this point, fuel is injected into the combustion chamber and is ignited by the heat of the compression. This begins the power stroke. The expanding force of the burning gases pushes the piston downward, providing power to the crankshaft. The diesel fuel will continue to burn through the entire power stroke (a more complete burning of the fuel). The gasoline engine has a power stroke with rapid combustion in the beginning, but little to no combustion at the end (Fig. 1.14(c)).

Exhaust Stroke: As the piston reaches bottom dead center on the power stroke, the power stroke ends and the exhaust stroke begins. The exhaust valve opens, and, as the piston rises towards top dead center, the burnt gases are pushed out through the exhaust port. As the piston reaches top dead center, the exhaust valve closes and the intake valve opens. The engine is now ready to begin another operating cycle (Fig. 1.14(d)).

Working of the four stroke CI engine cycle on p-V diagram as shown in Fig. 1.15.

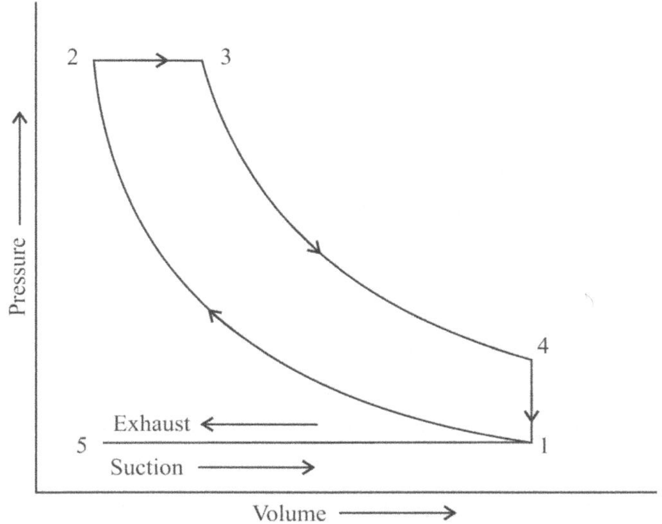

Fig. 1.15 Theoretical p-V Diagram of a Four Stroke Diesel Cycle Engine

Process 5-1: In this process fresh air is passed inside the cylinder. The piston moves from top dead center to bottom dead center. This comprises the first stroke of the engine.

Process 1-2: In this process the air is compressed and piston is moved to top dead center. This comprises the second stroke of the engine.

Process 2-3: In this process diesel is pumped into the cylinder through fuel pump and the fuel is burnt. This process is of constant volume and increase in pressure.

Process 3-4: In this process the burnt fuel expands itself and exerts pressure on the piston. The piston moves from top dead center to bottom dead center. This comprises the third stroke of the engine and the power stroke.

Process 4-1: In this process the burnt gas is exhausted out and the pressure decreases with constant volume.

Process 1-5: In this process the burnt gas is completely moved out of the cylinder by the action of piston. Piston moves from bottom dead center to top dead center. This comprises the fourth stroke of the engine.

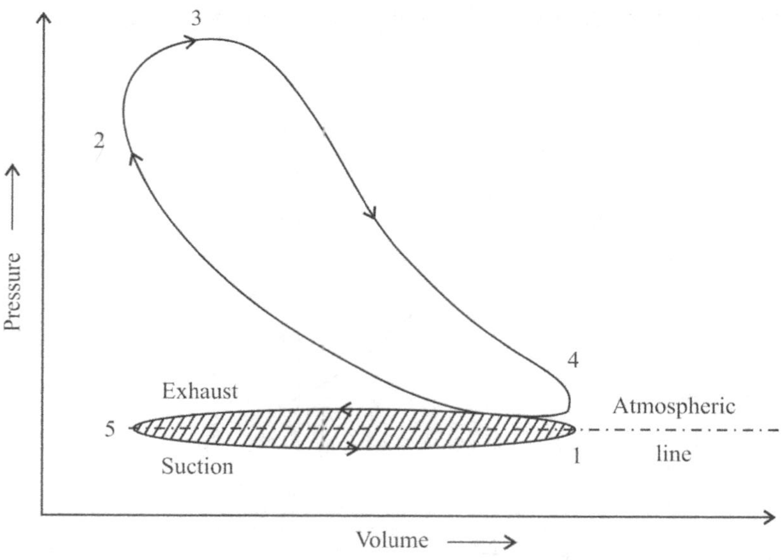

Fig. 1.16 Actual P-V Diagram of a Four Stroke Diesel Cycle Engine

The actual indicator diagram of 4-stroke cycle diesel engine has been shown in Fig. 1.16. The suction stroke is shown by line 1-2 which lies below the atmospheric pressure line. This pressure difference makes the fresh air to flow into the engine cylinder. The inlet valve offers some resistance to the incoming air. That is why, the air can not enter suddenly into the engine cylinder. As a result of this, pressure inside the cylinder remains somewhat below the atmospheric pressure during the suction stroke. The compression stroke is shown by the line 2-3, which shows that the inlet valve closes a little beyond 2 (i.e., B.D.C.). At the end of this stroke, there is an increase of pressure inside the engine cylinder, shortly before the end of the compression stroke (i.e., T.D.C.), fuel valve opens and the fuel is injected into the engine cylinder. The fuel is ignited by the temperature of the compressed air. The ignition suddenly increases volume and temperature of the products of combustion. But the pressure, practically remains constant. The expansion stroke is shown by the line 4-5 in which the exit valve opens a little before 5 (i.e., B.D.C.). Now the burnt gases are exhausted into the atmosphere through the exhaust valve, shown by 5-1, lies above atmospheric pressure line. The exhaust valve offers some resistance to the outgoing burnt gases. That is why, the burnt gases can not escape suddenly from the engine cylinder. As a result of this pressure inside the cylinder remains above the atmospheric pressure during the exhaust stroke.

1.7   Comparison of Spark Ignition (S.I.) and Compression Ignition (C.I.) Engines

The most prominent difference between Spark Ignition (SI) and Compression Ignition (CI) engines is the type of fuel used in each. In SI engines petrol or gasoline is used as fuel, hence these engines are also called petrol engines. In CI engines diesel is used as fuel, hence they are also called diesel engines. Here are some other major differences between the SI and CI engines is given in Table 1.2.

Table 1.2 Comparison of Spark Ignition (S.I.) and
Compression Ignition (C.I.) Engines

S. No	Features	S.I. Engines	C.I. Engines
1.	Thermodynamic Cycle	Otto cycle	Diesel cycle for slow speed engines. Dual cycle for high speed engines.
2.	Fuel used	Petrol (Gasoline)	Diesel
3.	Air fuel (A/F) ratio	10:1 to 20:1	18:1 to 100:1
4.	Compression Ratio	Up to 11; Average value 7 to 9 Upper limit fixed by anti-knock quality of fuel.	Up to12 to 24; Average value 15 to 18; Upper limit is limited by thermal and mechanical stresses.

Table 1.2 *Contd...*

S. No	Features	S.I. Engines	C.I. Engines
5.	Combustion	Spark Ignition	Compression Ignition
6.	Fuel Supply	By Carburetor (Low Cost)	By injection (High Cost)
7.	Operating pressure 1. Compression pressure 2. Maximum Pressure	7 bar to 15 bar 45 bar to 60 bar	30 bar to 50 bar 60 bar to 120 bar
8.	Operating Speed	High speed: 2000 to 6000 r.p.m	Low speed: 400 r.p.m Medium speed: 400 to 1200 r.p.m. High speed: 1200 to 3500 r.p.m
9.	Power Control	Throttle governed the quantity	Rack governed the quantity
10.	Calorific value	44 MJ/kg	42 MJ/kg
11.	Cost of running	High	Low
12.	Maintenance cost	Minor maintenance	Major over all require
13.	Super charging	Limited by detonation. Used only in aircraft engines.	Limited by lower power and mechanical and thermal stresses. Widely used.
14.	Two stroke operation	Less suitable, fuel loss in scavenging. But small two stroke engines are used in mopeds, scooters and motor cycles due to their simplicity and low cost.	No fuel loss in scavenging. More suitable.
15.	High powers	No	Yes
16.	Distribution of fuel	A/F ratio is not optimum in multi cylinder engines.	Excellent distribution of fuel in multi cylinder engines.
17.	Starting	Easy, low cranking effort.	Difficult, high cranking effort.
18.	Exhaust gas temperature.	High, due to low thermal efficiency	Low, due to high thermal efficiency.
19.	Weight per unit power	Low (0.5 to 4.5 kg/kW)	High (3.3 to 13.5 kg/kW)

Table 1.2 Contd...

S. No	Features	S.I. Engines	C.I. Engines
20.	Initial capital Cost	Low	High due to heavy weight and study construction; costly construction 1.25-1.5 times.
21.	Noise and distribution	Less	More idle noise problem
22.	Applications	Mopeds, Scooters, motorcycles, Simple engine passenger cars	Buses, trucks, locomotives, tractors, earth moving machinery and stationary generating plants.

1.8 Comparison between 2-Stroke and 4-Stroke Engines

The two-stroke engine was developed to obtain a greater output from the same size of the engine. The engine mechanism also eliminates the valve arrangement making it mechanically simpler. Almost all two-stroke engines have no conventional valves but only ports (some have an exhaust valve). This simplicity of the two-stroke engine makes it cheaper to produce and easy to maintain. Theoretically a two-stroke engine develops twice the power of a comparable four stroke engine because of one power stroke every revolution (compared to one power stroke every two revolutions of a four-stroke engine). This makes the two-stroke engine more compact than a comparable four-stroke engine. In actual practice power output is not exactly doubled but increased by only about 30% because of

(i)   Reduced effective expansion stroke and

(ii) Increased heating caused by increased number of power strokes that limits the maximum speed.

The other advantages of the two-stroke engines are more uniform torque on crankshaft and comparatively less exhaust gas dilution. However, when applied to the spark-ignition engine the two stroke cycle has certain disadvantages which have restricted its application to only small engines suitable for motor cycles, scooters, lawn mowers, outboard engines etc. In the SI engine, the incoming charge consists of fuel and air. During scavenging, as both inlet and exhaust ports are open simultaneously for some time, there is a possibility that some of the fresh charge containing fuel escapes with the exhaust. This results in high fuel consumption and lower thermal efficiency. The other drawback of two-stroke engine is the lack of flexibility, viz., the capacity to operate with the same efficiency at all speeds. At part throttle operating condition, the amount of fresh mixture entering the cylinder is not enough to clear all the exhaust gases and a part of it remains in the cylinder to contaminate the charge. This results in irregular operation of the engine. The two-stroke diesel engine does not suffer from these defects. There is no loss of fuel with exhaust gases as the intake charge in diesel engine is only air. The two-stroke diesel engine is used quite widely. Many of the high output diesel engines work on this cycle. A disadvantage common to all two-stroke engines, gasoline as well as diesel, is the greater cooling and lubricating oil requirements due to one power stroke in each revolution of the crankshaft. Consumption of lubricating oil is high in two-stroke engines due to higher temperature. A detailed comparison of two-stroke and four-stroke engines are given in the Table below.

Table 1.3 Comparison of two-stroke and four-stroke engines

S.No	Features	Four Stroke Cycle Engines	Two Stroke Cycle Engines
1.	Completion of cycle	The cycle is completed in four strokes of the piston or in two revolutions of the crank shaft. Thus one power stroke is obtained in every two revolutions of the crank shaft.	The cycle is completed in two strokes of the piston or in one revolution of the crank shaft. Thus power stroke is obtained in each revolutions of the crank shaft.
2.	Flywheel required heavier or lighter	Because of the turning movement is not so uniform and hence heavier flywheel is needed.	More uniform turning movements and hence lighter flywheel is needed.
3.	Power produced for same size of engine	Again because of one power stroke for two revolutions, power produced for same size of engine is small or for the same power the engine is heavy and bulky.	Because of one power stroke for one revolution. power produced for same size of engine is more (Theoretically twice, actually about 1.3 times) or for the same power the engine is light and compact.

Table 1.3 *Contd...*

S.No	Features	Four Stroke Cycle Engines	Two Stroke Cycle Engines
4.	Cooling and lubrication requirements	Because of one power stroke in two revolutions lesser cooling and lubrication requirements. Lesser rate of wear and tear.	Because of one power stroke in one revolutions greater cooling and lubrication requirement. Large rate of wear and tear.
5.	Valve and valve mechanism	The four stroke engine contains valve and valve mechanism.	Two stroke engines have no valves but only ports (some two stroke engines are fitted with conventional exhaust valves).
6.	Initial cost	Because of heavy weight and complication of valve mechanism, the initial cost is higher.	Because of light weight and simplicity due to absence of valves mechanism, cheaper in initial cost.
7.	Volumetric efficiency	Volumetric efficiency more due to more time of induction.	Volumetric efficiency less due to lesser time for induction.
8.	Thermal and part load efficiency	Thermal efficiency higher, part load efficiency better than two stroke cycle engine.	Thermal efficiency lower, part load efficiency lesser than four stroke cylinder engine.
9.	Applications	Used where efficiency is important; in cars, buses, trucks, tractors, industrial engines, aeroplane, power generators etc.	In two strokes petrol engine some fuel is exhausted during scavenging. Used where (a) low cost, and (b) compactness and light weight important. Two stroke (air cooled) petrol engines used in very small sizes only, lawn movers, scooters, motor cycles (lubrication oil mixed with petrol). Two stroke diesel engines used in very large sizes more than 60 cm bore, for ship propulsion because of low weight and compactness.

1.9 Thermodynamic Cycles

1.9.1 Carnot Cycle

Nicolas Leonard Sadi Carnot was the first to provide a thermodynamic model of a heat engine, abstracting from the only available heat engine, the steam engine, to pinpoint the fundamentals: the idea of a generic working fluid, performing a generic cyclic process, interacting with generic heat reservoirs. In that his only publication, Carnot concluded that all heat engines where limited in their energy-conversion efficiency by the operating temperatures, and that the maximum efficiency is obtained when the working fluid is assumed to follow four ideal processes (Fig. 1.17):

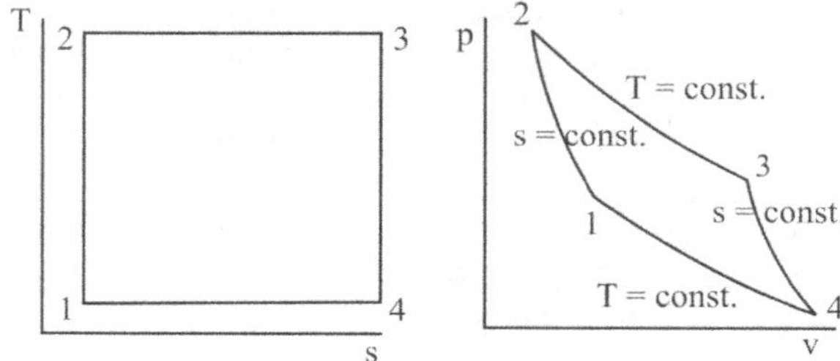

Fig. 1.17 The Carnot Cycle in T-s and p-V Diagrams

- An isentropic compression (1 to 2), to change temperature without heat transfer.
- An isothermal heat input (2 to 3), from the hot source, at the hot-source temperature.
- An isentropic expansion (3 to 4), to change temperature without heat transfer.
- An isothermal heat rejection to the cold source (usually the environment), at the cold-source temperature.

Carnot reached those conclusion by a set of rational deductions, namely: any engine with friction would have less efficiency than one without, among all engines exchanging heat at different temperatures, the one with highest efficiency only exchanges heat at the two extreme temperatures (the hottest and the coldest), and all reversible engines working with the same couple of temperature extremes have the same efficiency.

The energy can be easily deduced by establishing the overall energy conservation, $\Delta E_{univ} = Q_{hot}WQ_{cold} = 0$, and the overall entropy balance, $\Delta S_{uni}Q_{hot}/T_{hot}+Q_{cold}/T_{cold}, \geq 0$, the latter being zero in the ideal case of a Carnot cycle, what yields:

$$\eta_e \equiv \frac{W_{net}}{Q_{pos}} \rightarrow \eta_{e.Carrot} = 1 - \frac{T_1}{T_2}$$

Carnot cycle, and any other conceivable power cycle, is to be run clockwise in both the T-s and p-V thermodynamic diagrams, since the heat engine receives net heat, that in a reversible process is $Q_{net} = \int TdS$ (recall the area interpretation of integrals), and delivers net work, that in a reversible process is $W_{net} = \int pdV$; notice that in a cycle $\int dU = \int Td\tilde{S} \int pdV = 0$.

The Carnot cycle is not practical, not only because of unavoidable frictional losses (could be minimised with appropriate lubrication), but because of the heat transfer with negligible temperature jump, that would render the heat transfer rate infinitesimal for a finite size engine with finite thermal transmittance with the heat sources.

1.9.2 Otto Cycle

The Otto cycle is a first approximation to model the operation of a spark-ignition engine, first built by Nikolaus Otto in 1876, and used in many cars, small planes and small power systems.

In the ideal air-standard Otto cycle, the working fluid is just air, which is assumed to follow four processes: isentropic compression, constant-volume heat input from the hot source, isentropic expansion, and constant-volume heat rejection to the environment (Fig. 1.18).

The main parameters of ideal and real Otto cycles are:
(i) Size, measured by the displacement volume (the volume swept by the piston, V_1-V_2), usually less than 0.5 litres per cylinder, to avoid self-ignition.
(ii) Speed, more precisely crankshaft speed, n, with a typical operation range n = 1000-7000 rpm (n = 20-120 Hz). The maximum value may be in the range n_{max}= 6000-8000 rpm for four-stroke engines. Two-stroke motorcycle engines run quicker (n_{max} = 13 000 rpm), the quickest (n_{max} = 20 000 rpm) being the smallest engines (two stroke), used in aircraft modelling.

(iii) Compression ratio, $r = V_1/V_2$, with a typical range of $r = 8\text{-}10$ (up to 14 in direct-injection spark-ignition engines), limited by the knock or self-ignition problem.

(iv) Mean effective pressure, p_{mep}, defined as the unit work divided by the displacement, with a typical range of 0.2-1.5 MPa (the full-load value may range from $p_{mep} = 1.2$ MPa in two-stroke motorcycle engines, to $p_{mep} = 1.7$ MPa in the largest turbocharged engines). Maximum pressure may have a typical range of 4-10 MPa. Performance maps of reciprocating engines are usually presented on p_{mep} Vs N diagram, i.e., mean effective pressure versus engine speed.

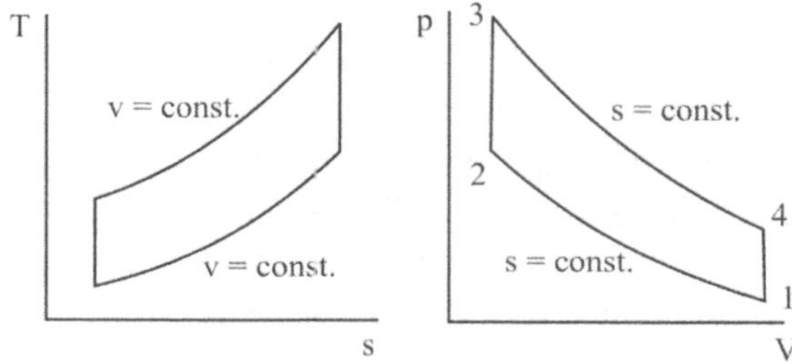

Fig. 1.18 The Ideal Otto Cycle in the T-s and p-V Diagram

The cold-air-standard model takes as working fluid air with constant properties (those at the inlet, i.e. cold), what renders the analysis simple. The energy exchanges for the trapped control mass, m, are $W_{12}/m = c_v(T_2 - T_1)$, $Q_{12} = 0$, $W_{23} = 0$, $Q_{23}/m = c_v(T_3 - T_2)$, $W_{34}/m = c_v(T_4 - T_3)$, $Q_{34} = 0$, $W_{41} = 0$, $Q_{41}/m = c_v(T_1 - T_4)$, and the energy efficiencies is:

$$\eta_{e,Otto} = \frac{W_{34} - W_{12}}{Q_{23}} = \frac{(T_3 - T_4) - (T_2 - T_1)}{T_3 - T_2} = 1 - \frac{T_1}{T_2} = 1 - \frac{1}{r^{\gamma-1}}$$

where, $\gamma = c_p/c_v$, 1.4 approx (for air)

Typical energy efficiencies of these engines are low, 25% to 35% when running at nominal power, and much lower at partial load (the main air flow is strangulated), but the engine is light, very powerful and responsive (accelerates very quickly).

One of the key advantages of Otto engines, is that cheap materials can be used in their construction (e.g., cast iron, against very expensive nickel alloys in gas turbines), because peak temperatures (up to 3000 K) are only realised within the burning gases and for short times, with an average temperature during the whole cycle of some 700 K, which would be the quasi-steady temperature level at the wall. Pollutant emission is higher because premixed fuel cannot burn close to the walls (the effect of walls is smaller in larger diesel engines, and most of the nearby gas is just non-premixed air), making the exhaust catalyser an environmental need.

1.9.3 Diesel Cycle

The Diesel cycle is a first approximation to model the operation of a compression-ignition engine, first built by Rudolf Diesel in 1893, and used in most cars, nearly all trucks, nearly all boats, many locomotives, some small airplanes, and many large electric power systems and cogeneration systems. It is the reference engine from 50 kW to 50 MW, due to the fuel used (cheaper and safer than gasoline) and the higher efficiency.

In the ideal air-standard Diesel cycle, the working fluid is just air, which is assumed to follow four processes (Fig. 1.19): isentropic compression, constant-pressure heat input from the hot source, isentropic expansion, and constant-volume heat rejection to the environment.

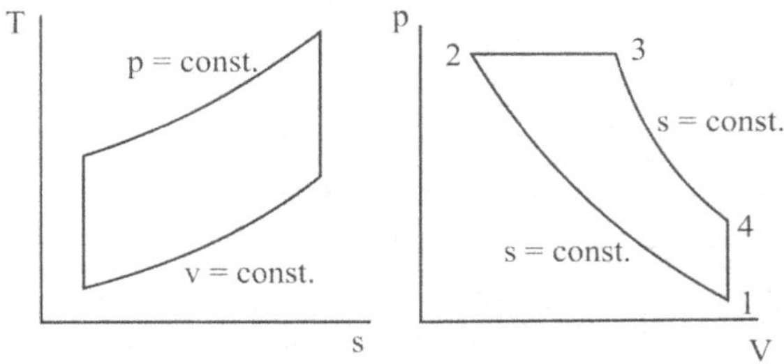

Fig. 1.19 The Ideal Otto Cycle in the T-s and p-V Diagram

Similarly to the Otto cycle, the main parameters of ideal and real Diesel cycles are also the size, measured by the displacement volume (that may reach more than 1 m^3 per cylinder in large marine engines), the compression ratio, $r = V_1/V_2$ (with a typical range of 16-22, limited just by strength), the cut-off ratio, r_c, or the mean effective pressure (in the range 1-2 MPa), or the maximum pressure (in the range of 3 MPa to 20 MPa), and the speed (with a typical range of 100-6000 rpm). The energy efficiency can be expressed as:

$$\eta_{e,\text{Diesel}} = 1 - \frac{1}{r^{\gamma-1}} \frac{r_c^{\gamma} - 1}{\gamma(r_c - 1)}$$

Typical energy efficiencies are from 30% to 54% (based on the lower heating value of the fuel). Real compression-ignition engines take ambient air (often after a first stage compression) and compress it (inside the cylinder) so much, rising the temperature accordingly, that the fuel burns as it is injected (after a small initial delay due to vaporisation and combustion kinetics).

One of its key advantages compared to Otto engines is the great load increase per cylinder associated to the higher pressures allowed (the mixture of fuel and air would deonate in Otto engines at high compressions), and the further load increase associated to charging previously-compressed air (turbocharging). Another advantage is the much better performance at part load, since there it is achieved by injecting less fuel instead of by throttling, and the torque (power divided per angular speed), changes less with angular speed.

Table 1.4 Comparison between Otto and Diesel Cycle

Otto Cycle	Diesel Cycle
Petrol engine works on Otto cycle	Diesel engine works on Diesel cycle
During suction stroke, air petrol mixture is sucked into the cylinder.	During suction stroke, only air is sucked into the cylinder.
Heat is added at constant volume.	Heat is added at constant pressure.
Efficiency is higher, for the same compression ratio	Efficiency is low, for the same compression ratio
Compression ratio has to be kept below 12 due to knocking	No limitation
Thermal efficiency is lower due to lower CR.	Thermal efficiency is higher due to higher CR.

Review Questions

1. Draw the Otto cycle on p-V diagram.

2. What is carnot cycle and its importance?

3. What is indicator diagram?

4. Why scavenging is important in two stroke engines compared to four stroke engines

5. Write any four advantage of 4 stroke petrol engine over 2 stroke petrol engine.

6. Write the air standard efficiency of an Otto cycle.

7. Compare four stroke and two stroke cycle engines and bring out their relative merits and demerits.

8. Explain with neat sketches the two different types of two stroke engines

9. Discuss the two stroke engine construction and operation with neat sketches

10. Briefly explain the principle of four stroke petrol engine with a simple sketch

11. Draw the ideal and actual indicator diagram for a four stroke petrol engine and explain why they differ each other.

12. What are the advantage and disadvantage of 2 stroke engine?

13. What is thermodynamic cycle?

14. What are the assumptions made for air standard cycle analysis?

15. Mention the various process of diesel cycle.

16. Sketch Otto cycle on p-V diagram and name all the process.

17. Plot and explain briefly the Diesel cycle on p-V & T-s diagram.

18. Write down the comparison between the S.I engine and C.I engine.

19. Define the following terms.

    1. Compression ratio

    2. Cut off ratio

    3. Expansion ratio.

20. Name the factors that affect air standard efficiency of Diesel cycle?

21. For the same compression ratio and heat supplied, state the order of decreasing air standard efficiency of Otto and diesel.

22. What is the range of compression ratio for Otto and Diesel cycle?

23. Differentiate between the auto cycle and diesel cycle.

CHAPTER 2

Internal Combustion Engine System

2.1 Ignition Systems

2.1.1 Introduction

We know that in case of Internal Combustion (IC) engines, combustion of air and fuel takes place inside the engine cylinder and the products of combustion expand to produce reciprocating motion of the piston. This reciprocating motion of the piston is in turn converted into rotary motion of the crank shaft through connecting rod and crank. This rotary motion of the crank shaft is in turn used to drive the generators for generating power. We also know that there are 4-cycles of operations viz.: suction; compression; power generation and exhaust. These operations are performed either during the 2-strokes of piston or during 4-strokes of the piston and accordingly they are called as 2-stroke cycle engines and 4-stroke cycle engines. In case of petrol engines during suction operation, charge of air and petrol fuel will be taken in. During compression this charge is compressed by the upward moving piston. And just before the end of compression, the charge of air and petrol fuel will be ignited by means of the spark produced by means of spark plug and the ignition system does the function of producing the spark in case of spark ignition engines.

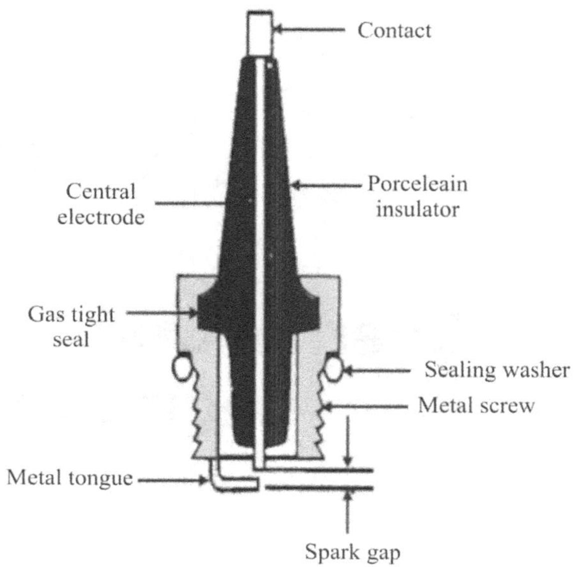

Fig. 2.1 Spark Plug

Fig. 2.1 shows a typical spark plug used with petrol engines. It mainly consists of a central electrode and metal tongue. Central electrode is covered by means of porcelain insulating material. Through the metal screw the spark plug is fitted in the cylinder headplug. When the high tension voltage of the order of 30000 volts is applied across the spark electrodes, current jumps from one electrode to another producing a spark. Whereas in case of diesel (Compression Ignition-CI) engines only air is taken in during suction operation and is compressed during compression operation and just before the end of compression, when diesel fuel is injected, it gets ignited due to heat of compression of air. Once the charge is ignited, combustion starts and products of combustion expand, i.e. they force the piston to move downwards i.e., they produce power and after producing the power the gases are exhausted during exhaust operation.

Objectives

After studying this chapter, you should be able to

- explain the different types of ignition systems,
- differentiate between battery and magneto ignition system,
- know the drawbacks of conventional ignition system, and
- appreciate the importance of ignition timing and ignition advance.

2.1.2 Ignition System Types

Basically Convectional Ignition systems are of two types:

    (a) Battery or Coil Ignition System, and

    (b) Magneto Ignition System.

Both these conventional ignition systems work on mutual electromagnetic induction principle. Battery ignition system was generally used in 4-wheelers, but now-a-days it is more commonly used in 2-wheelers also (i.e., Button start, 2-wheelers like Pulsar, Kinetic Honda; Honda-Activa, Scooty, Fiero, etc.,). In this case 6 V or 12 V batteries will supply necessary current in the primary winding. Magneto ignition system is mainly used in 2-wheelers, kick start engines.(Example, Bajaj Scooters, Boxer, Victor, Splendor, Passion, etc.,). In this case magneto will produce and supply current to the primary winding. So in magneto ignition system magneto replaces the battery.

(a) **Battery or Coil Ignition System:** Fig. 2.2 shows line diagram of battery ignition system for a 4-cylinder petrol engine. It mainly consists of a 6 or 12 volt battery, ammeter, ignition switch, auto-transformer (step up transformer), contact breaker, capacitor, distributor rotor, distributor contact points, spark plugs, etc. Here there are 4-spark plugs and contact breaker cam has 4-corners. (If it is for 6-cylinder engine it will have 6-spark plugs and contact breaker cam will be a perfect hexagon).

    The ignition system is divided into 2-circuits:

       (i) *Primary Circuit:* It consists of 6 or 12 V battery, ammeter, ignition switch, primary winding it has 200-300 turns of 20 SWG (Sharps Wire Gauge) gauge wire, contact breaker, capacitor.

      (ii) *Secondary Circuit:* It consists of secondary winding. Secondary **Ignition Systems** winding consists of about 21000 turns of 40 (SWG) gauge wire. Bottom end of which is connected to bottom end of primary and top end of secondary winding is connected to centre of distributor rotor. Distributor rotors rotate and make contacts with contact points and are connected to spark plugs which are fitted in cylinder heads(engine earth).

(iii) ***Working:*** When the ignition switch is closed and engine in cranked, as soon as the contact breaker closes, a low voltage current will flow through the primary winding. It is also to be noted that the contact breaker cam opens and closes the circuit 4-times (for 4 cylinders) in one revolution. When the contact breaker opens the contact, the magnetic field begins to collapse. Because of this collapsing magneticfield, current will be induced in the secondary winding and because of more turns (@21000 turns) of secondary, voltage goes up to 28000-30000 volts.

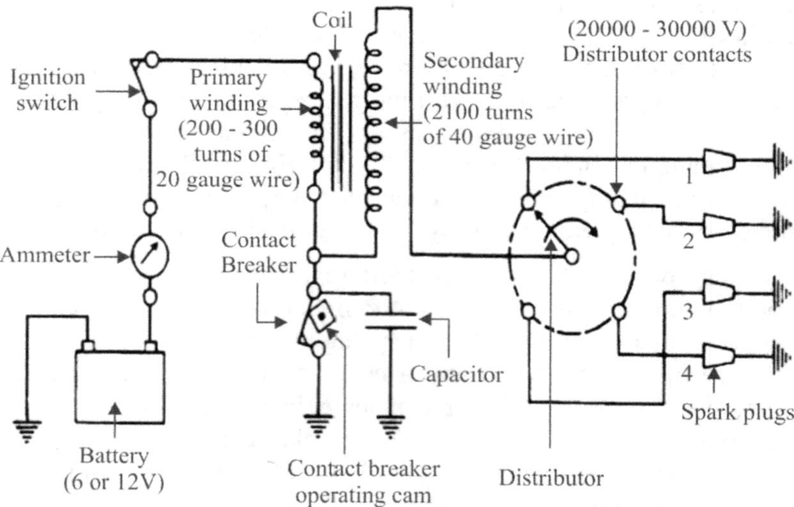

Fig. 2.2 Schematic Diagram of Coil/Battery Ignition System

This high voltage current is brought to centre of the distributor rotor. Distributor rotor rotates and supplies this high voltage current to proper spark plug depending upon the engine firing order. When the high voltage current jumps the spark plug gap, it produces the spark and the charge is ignited-combustion starts-products of combustion expand and produce power.

Note:

(a) The function of the capacitor is to reduce arcing at the contact breaker (CB) points. Also when the CB opens the magnetic field in the primary winding begins to collapse. When the magnetic field is collapsing capacitor gets fully charged and then it starts discharging and helps in building up of voltage in secondary winding.

(b) Contact breaker cam and distributor rotor are mounted on the same shaft. In 2-stroke cycle engines these are motored at the same engine speed. And in 4-stroke cycle engines they are motored at half the engine speed.

(b) **Magneto Ignition System:** In this case magneto will produce and supply the required current to the primary winding. In this case as shown, we can have rotating magneto with fixed coil or rotating coil with fixed magneto for producing and supplying current to primary, remaining arrangement is same as that of a battery ignition system. Figure 2.3 shows the line diagram of magneto ignition system

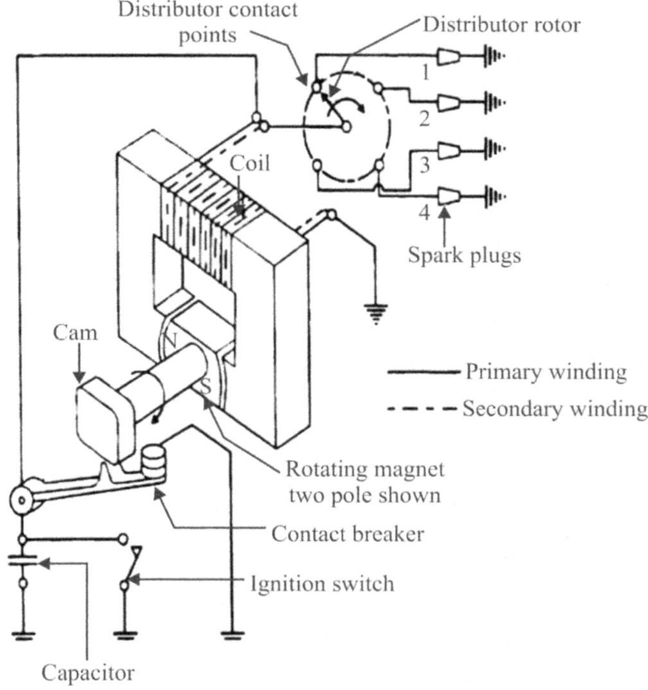

Fig. 2.3 Schematic Diagram of Magneto Ignition System

2.1.3 Comparison between Battery and Magneto Ignition Systems

Battery Ignition	Magneto Ignition
Battery supplies current in primary circuit	No battery needed
A good spark is available at low speed also	Magneto produces the required current for primary circuit
Occupies more space	During starting the quality of spark is poor due to slow speed
Recharging is a must in case battery gets discharged	No such arrangement required
Mostly employed in car and bus for which it is required to crank the engine	Used on motorcycles, scooters, etc
Battery maintenance is required	No battery maintenance problems

2.1.4 Drawbacks (Disadvantages) of Conventional Ignition Systems

Following are the drawbacks of conventional ignition systems:

(a) Because of arcing, pitting of contact breaker point and which will lead to regular maintenance problems.

(b) *Poor starting:* After few thousands of kilometres of running, the timing becomes inaccurate, which results into poor starting (Starting trouble).

(c) At very high engine speed, performance is poor because of inertia effects of the moving parts in the system.

(d) Sometimes it is not possible to produce spark properly in fouled spark Ignition Systems plugs.

In order to overcome these drawbacks Electronic Ignition system is used.

2.1.5 Advantages of Electronic Ignition System

Following are the advantages of electronic ignition system:

(a) Moving parts are absent – so no maintenance.

(b) Contact breaker points are absent – so no arcing.

(c) Spark plug life increases by 50% and they can be used for about 60000 km without any problem.

(d)   Better combustion in combustion chamber, about 90-95% of air fuel mixture is burnt compared with 70-75% with conventional ignition system.

(e)   More power output.

(f)   More fuel efficiency.

2.1.6 Types of Electronic Ignition System

Electronic Ignition System is as follow:

(a)   Capacitance Discharge Ignition system

(b)   Transistorized system

(c)   Piezo-electric Ignition system

(d)   The Texaco Ignition system

Capacitance Discharge Ignition System: Complete picture of capacitance discharge ignition system is shown in Fig. 2.4. It mainly consists of 6-12 V battery, ignition switch, DC to DC convertor, charging resistance, tank capacitor, Silicon Controlled Rectifier (SCR), SCR-triggering device, step up transformer, spark plugs. A 6-12 volt battery is connected to DC to DC converter i.e., power circuit through the ignition switch, which is designed to give or increase the voltage to 250-350 volts. This high voltage is used to charge the tank capacitor (or condenser) to this voltage through the charging resistance. The charging resistance is so designed that it controls the required current in the SCR.

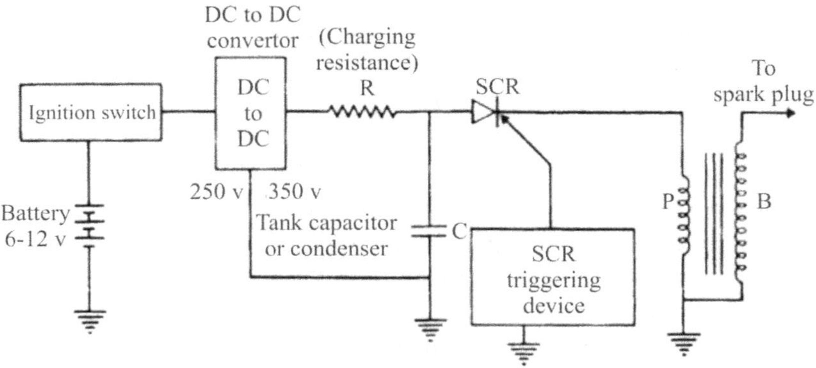

Fig. 2.4 Capacitance Discharge Ignition System

Depending upon the engine firing order, whenever the SCR triggering device, sends a pulse, then the current flowing through the primary winding is stopped and the magnetic field begins to collapse. This collapsing magnetic field will induce or step up high voltage current in the secondary, which while jumping the spark plug gap produces the spark, and the charge of air fuel mixture is ignited.

Transistorized Assisted Contact (TAC) Ignition System: Figure 2.5 shows the TAC system.

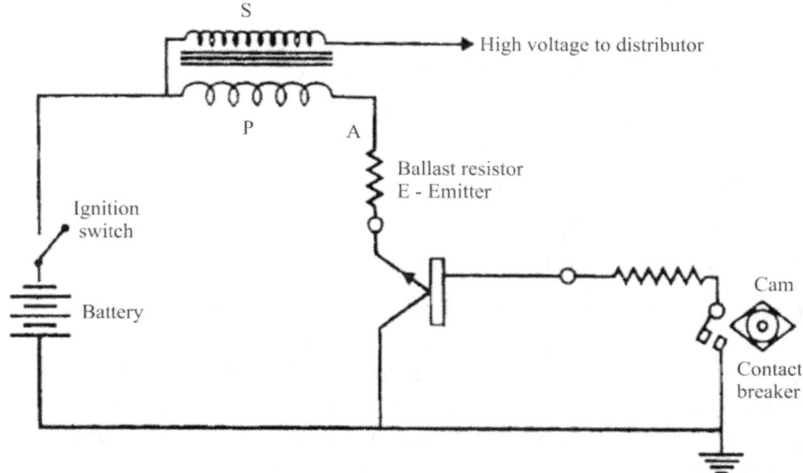

Fig. 2.5 Transistorized Assisted Contact (TAC) Ignition System

Advantages
  (a) The low breaker-current ensures longer life.
  (b) The smaller gap and lighter point assembly increase dwell time minimize contact bouncing and improve repeatability of secondary voltage.
  (c) The low primary inductance reduces primary inductance reduces primary current drop-off at high speeds.

Disadvantages
  (a) As in the conventional system, mechanical breaker points are necessary for timing the spark.
  (b) The cost of the ignition system is increased.
  (c) The voltage rise-time at the spark plug is about the same as before.

Piezo-electric Ignition System: The development of synthetic piezo-electric materials producing about 22 kV by mechanical loading of a small crystal resulted in some ignition systems for single cylinder engines. But due to difficulties of high mechanical loading need of the order of 500 kg timely control and ability to produce sufficient voltage, these systems have not been able to come up.

The Texaco Ignition System: Due to the increased emphasis on exhaust emission control, there has been a sudden interest in exhaust gas recirculation systems and lean fuel-air mixtures. To avoid the problems of burning of lean mixtures, the Texaco Ignition system has been developed. It provides a spark of controlled duration which means that the spark duration in crank angle degrees can be made constant at all engine speeds. It is an AC system. This system consists of three basic units, a power unit, a control unit and a distributor sensor.

This system can give stable ignition up to A/F ratios as high as 24: 1.

2.1.7 Firing Order

- The order or sequence in which the firing takes place, in different cylinders of a multicylinder engine is called Firing Order.

- In case of SI engines the distributor connects the spark plugs of different cylinders according to Engine Firing Order.

Advantages

(a) A proper firing order reduces engine vibrations.

(b) Maintains engine balancing.

(c) Secures an even flow of power.

Firing order differs from engine-to-engine.

Probable firing orders for different engines are:

$- 3$ cylinder $= 1 - 3 - 2$

$- 4$ cylinder engine (inline) $= 1 - 3 - 4 - 2$

$1 - 2 - 4 - 3$

$- 4$ cylinder horizontal opposed engine $= 1 - 4 - 3 - 2$

(Volkswagen engine)

$- 6$-cylinder in line engine $= 1 - 5 - 3 - 6 - 2 - 4$

(Cranks in 3 pairs) $1 - 4 - 2 - 6 - 3 - 5$

$1 - 3 - 2 - 6 - 4 - 5$

$1 - 2 - 4 - 6 - 5 - 3$

$- 8$ cylinder in line engine $1 - 6 - 2 - 5 - 8 - 3 - 7 - 4$

$1 - 4 - 7 - 3 - 8 - 5 - 2 - 6$

8 cylinder V type $1 - 5 - 4 - 8 - 6 - 3 - 7 - 2$

$1 - 5 - 4 - 2 - 6 - 3 - 7 - 8$

$1 - 6 - 2 - 5 - 8 - 3 - 7 - 4$

$1 - 8 - 4 - 3 - 6 - 5 - 7 - 2$

Cylinder 1 is taken from front of inline and front right side in V engines.

2.1.8 Importance of Ignition Timing and Ignition Advance Systems

Ignition timing is very important, since the charge is to be ignited just before (few degrees before TDC) the end of compression, since when the charge is ignited, it will take some time to come to the required rate of burning.

Ignition Advance: The purpose of spark advance mechanism is to assure that under every condition of engine operation, ignition takes place at the most favourable instant in time i.e., most favourable from a standpoint of engine power, fuel economy and minimum exhaust dilution. By means of these mechanisms the advance angle is accurately set so that ignition occurs before TDC point of the piston. The engine speed and the engine loads are the control quantities required for the automatic adjustment of the ignition timing. Most of the engines are fitted with mechanisms which are Integral with the distributor and automatically regulate the optimum spark advance to account for change of speed and load. The two mechanisms used are:

(a)   Centrifugal advance mechanism, and

(b)   Vacuum advance mechanism.

(a) **Centrifugal Advance Mechanism:** The centrifugal advance mechanism controls the ignition timing for full-load operation. The adjustment mechanism is designed so that its operation results in the desired advance of the spark. The cam is mounted, movably, on the distributor shaft so that as the speed increases, the flyweights which are swung farther and farther outward, shaft the cam in the direction of shaft rotation. As a result, the cam lobes make contact with the breaker lever rubbing block somewhat earlier, thus shifting the ignition point in the early or advance direction. Depending on the speed of the engine, and therefore of the shaft, the weights are swung outward a greater or a lesser distance from the centre. They are then held in the extended position, in a state of equilibrium corresponding to the shifted timing angle, by a retaining spring which exactly balances the centrifugal force. The weights shift the cam either or a rolling contact or sliding contact basis; for this reasons we distinguish between the rolling contact type and the sliding contact type of centrifugal advance mechanism. The beginning of the timing adjustment in the range of low engine speeds and the continuous adjustment based on the full load curves are determined by the size of the weights by the shape of the contact mechanisms (rolling or sliding contact type), and by the retaining springs, all of which can be widely differing designs. The centrifugal force controlled cam is fitted with a lower limit stop for purposes of setting the beginning of the adjustment, and also with an upper limit stop to restrict the greatest possible full load adjustment. A typical sliding contact type centrifugal advance mechanism is shown in Figures 2.6(a) and (b).

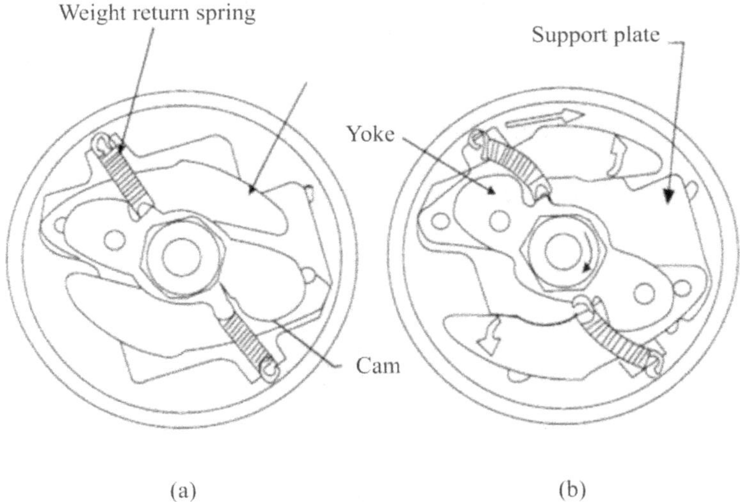

Fig. 2.6 Sliding Contact Type Centrifugal Advance Mechanism

(b) Vacuum Advance Mechanism: Vacuum advance mechanism shifts the ignition point under partial load operation. The adjustment system is designed so that its operation results in the prescribed partial load advance curve. In this mechanism the adjustment control quantity is the static vacuum prevailing in the carburettor, a pressure which depends on the position of the throttle valve at any given time and which is at a maximum when this valve is about half open. This explains the vacuum maximum. The diaphragm of a vacuum unit is moved by changes in gas pressure. The position of this diaphragm is determined by the pressure differential at any given moment between the prevailing vacuum and atmospheric pressure. The beginning of adjustment is set by the pre-established tension on a compression spring. The diaphragm area, the spring force, and the spring rigidity are all selected in accordance with the partial-load advance curve which is to be followed and are all balanced with respect to each other. The diaphragm movement is transmitted through a vacuum advance arm connected to the movable breaker plate, and this Support plat movement shifts the breaker plate an additional amount under partial load condition in a direction opposite to the direction of rotation of the distributor shaft. Limit stops on the vacuum advance arm in the base of the vacuum unit restrict the range of adjustment. The vacuum advance mechanism operates independent of the centrifugal advance mechanism. The mechanical interplay between the two advance mechanisms, however, permits the total adjustment angle at any given time to be the result of the addition of the shifts provided by the two individual mechanisms operates in conjunction with the engine is operating under partial load. A typical vacuum advance mechanism is shown in Figure 2.7.

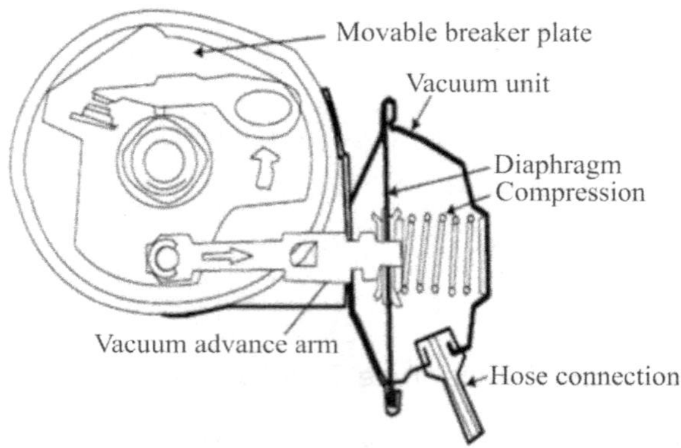

Fig. 2.7 Vacuum Advance Mechanism

2.2 Cooling Systems of IC Engines

2.2.1 Introduction

We know that in case of Internal Combustion engines, combustion of air and fuel takes place inside the engine cylinder and hot gases are generated. The temperature of gases will be around 2300-2500 °C. This is a very high temperature and may result into burning of oil film between the moving parts and may result into seizing or welding of the same. So, this temperature must be reduced to about 150-200 °C at which the engine will work most efficiently. Too much cooling is also not desirable since it reduces the thermal efficiency. So, the object of cooling system is to keep the engine running at its most efficient operating temperature.

It is to be noted that the engine is quite inefficient when it is cold and hence the cooling system is designed in such a way that it prevents cooling when the engine is warming up and till it attains to maximum efficient operating temperature, then it starts cooling.

It is also to be noted that:
 (a) About 20-25% of total heat generated is used for producing brake power (Useful work).
 (b) Cooling system is designed to remove 30-35% of total heat.
 (c) Remaining heat is lost in friction and carried away by exhaust gases.

Objectives

After studying this unit, you should be able to:
 • understand the methods of cooling of IC engine,
 • explain the air cooling system, and
 • know the water cooling system of IC engine.

2.2.2 Air Cooling System

There are mainly two types of cooling systems:
 (a) Air cooled system, and
 (b) Water cooled system.

 (a) **Air Cooled System:** Air cooled system is generally used in small engines say up to 15-20 kW and in aeroplane engines. In this system fins or extended surfaces are provided on the cylinder walls, cylinder head, etc (Fig. 2.8). Heat generated due to combustion in the engine cylinder will be conducted to the fins and when the air flows over the fins, heat will be dissipated to air.

The amount of heat dissipated to air depends upon:
 (a) Amount of air flowing through the fins.
 (b) Fin surface area.
 (c) Thermal conductivity of metal used for fins.

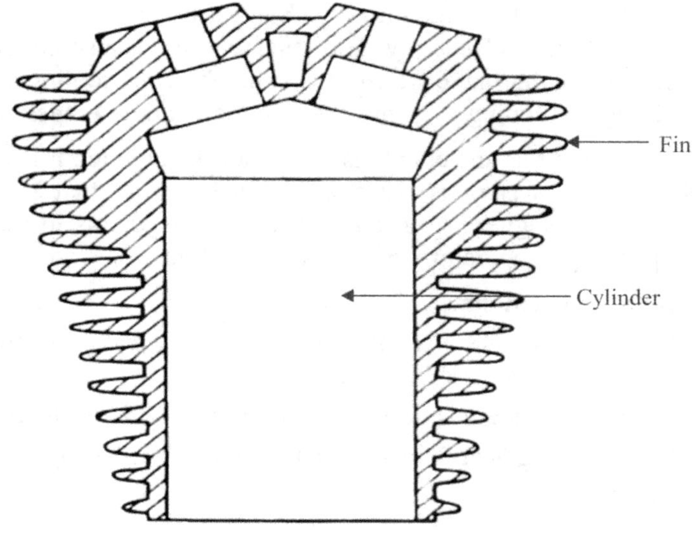

Fig. 2.8 Cylinder with Fins

Advantages of Air Cooled System

Following are the advantages of air cooled system:

(a) Radiator or pumps is absent hence the system is light.

(b) In case of water cooling system there are leakages, but in this case there are no leakages.

(c) Coolant and antifreeze solutions are not required.

(d) This system can be used in cold climates, where if water is used it may freeze.

Disadvantages of Air Cooled System

(a) Comparatively it is less efficient.

(b) It is used in aeroplanes and motorcycle engines where the engines are exposed to air directly.

2.2.3 Water Cooling System

In this method, cooling water jackets are provided around the cylinder, cylinder head, valve seats etc. The water when circulated through the jackets, it absorbs heat of combustion. This hot water will then be cooling in the radiator partially by a fan and partially by the flow developed by the forward motion of the vehicle. The cooled water is again recirculated through the water jackets.

Types of Water Cooling System: There are two types of water cooling system.

Thermo Siphon System: In this system the circulation of water is due to difference in temperature (i.e., difference in densities) of water as shown in Fig. 2.9. So in this system pump is not required but water is circulated because of density difference only.

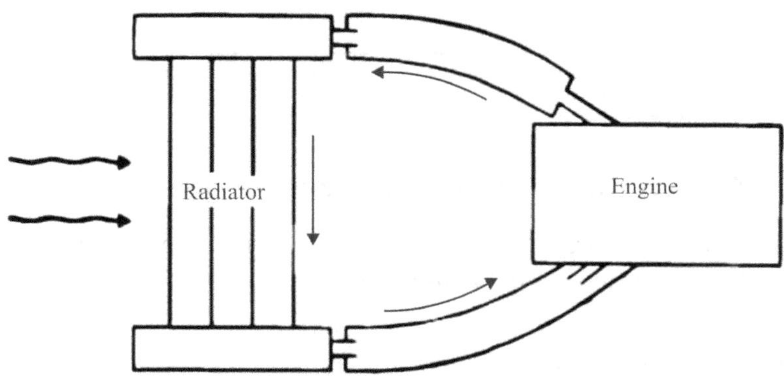

Fig. 2.9 Thermo Siphon System of Cooling

Pump Circulation System: In this system circulation of water is obtained by a pump (Fig. 2.10). This pump is driven by means of engine output shaft through V-belts.

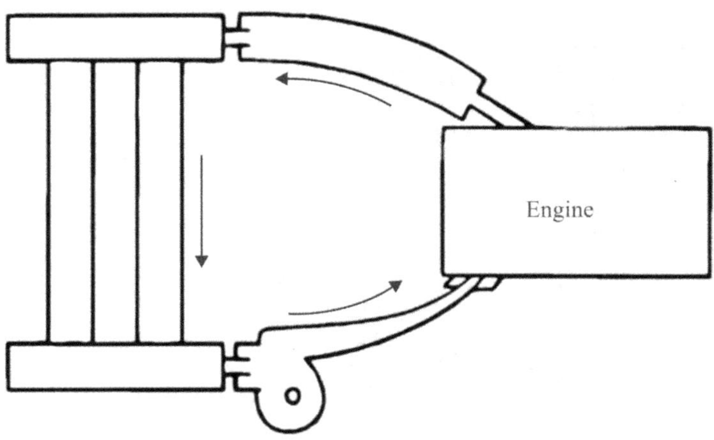

Fig. 2.10 Pump Circulation System

Components of Water Cooling System: Fig. 2.11 shows the various parts of water cooling system.

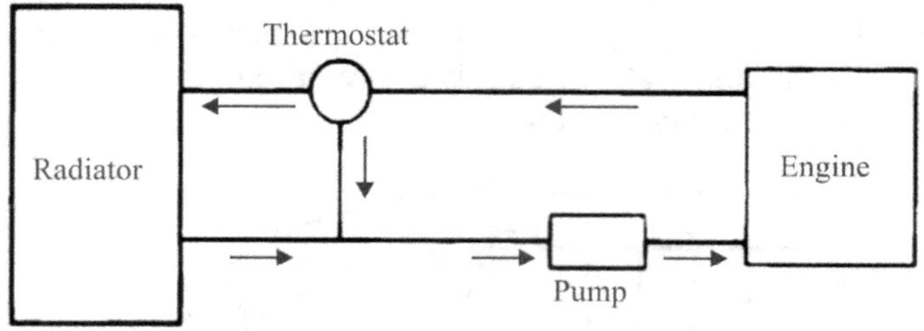

Fig. 2.11 Water Cooling System using Thermostat Valve

Figure 2.12 illustrate the water cooling system for a four cylinder engine. Water cooling system mainly consists of:

    (a) Radiator,

    (b) Thermostat valve,

    (c) Water pump,

    (d) Fan,

    (e) Water Jackets, and

    (f) Antifreeze mixtures.

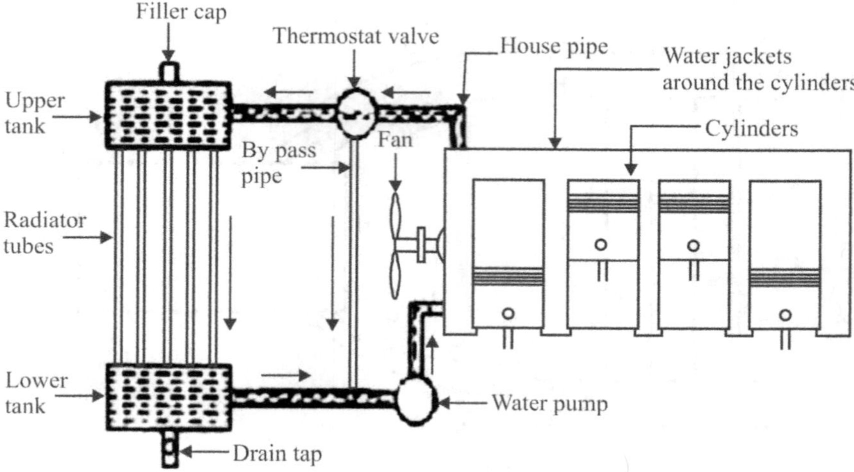

Fig. 2.12 Water Cooling System of a 4-cylinder Engine

Radiator: It mainly consists of an upper tank and lower tank and between them is a core. The upper tank is connected to the water outlets from the engines jackets by a hose pipe and the lover tank is connected to the jacket inlet through water pump by means of hose pipes.

There are 2-types of cores:

(a) Tabular

(b) Circular as shown in Fig. 2.13.

When the water is flowing down through the radiator core, it is cooled partially by the fan which blows air and partially by the air flow developed by the forward motion of the vehicle. As shown through water passages and air passages, water and air will be flowing for cooling purpose. It is to be noted that radiators are generally made out of copper and brass and their joints are made by soldering.

Thermostat Valve: It is a valve which prevents flow of water from the engine to radiator, so that engine readily reaches to its maximum efficient operating temperature. After attaining maximum efficient operating temperature, it automatically begins functioning. Generally, it prevents the water below 70 °C.

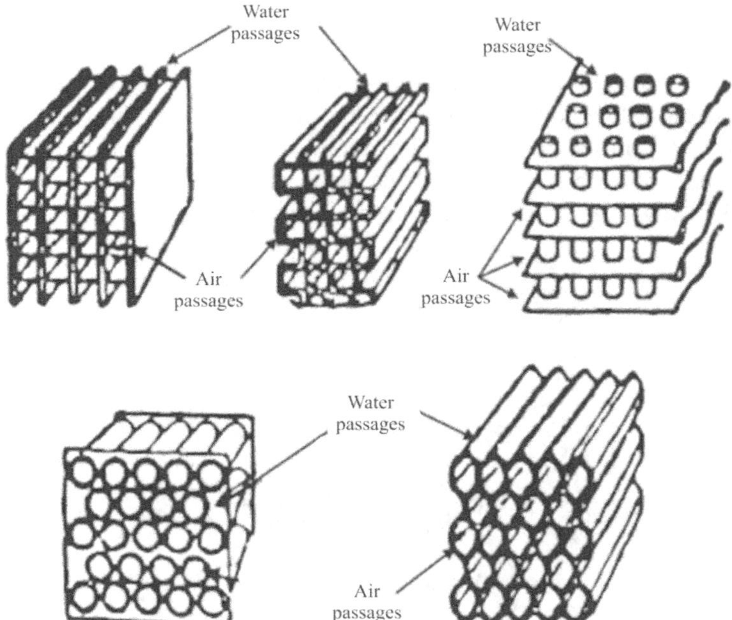

Fig. 2.13 Types of Cores (a) Tabular Radiator Sections and (b) Circular Radiator Sections

Fig. 2.14 shows the Bellow type thermostat valve which is generally used. It contains a bronze bellow containing liquid alcohol. Below is connected to the butterfly valve disc through the link. When the temperature of water increases, the liquid alcohol evaporates and the bellow expands and in turn opens the butterfly valve, and allows hot water to the radiator, where it is cooled.

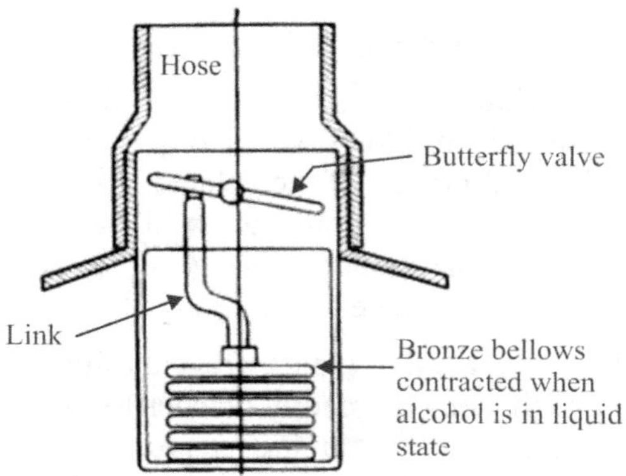

Fig. 2.14 Thermostat Valve

Water Pump: It is used to pump the circulating water. Impeller type pump will be mounted at the front end. Pump consists of an impeller mounted on a shaft and enclosed in the pump casing. The pump casing has inlet and outlet openings. The pump is driven by means of engine output shaft only through belts. When it is driven water will be pumped. Fig. 2.15 shows the cross sectioal view of water Pump.

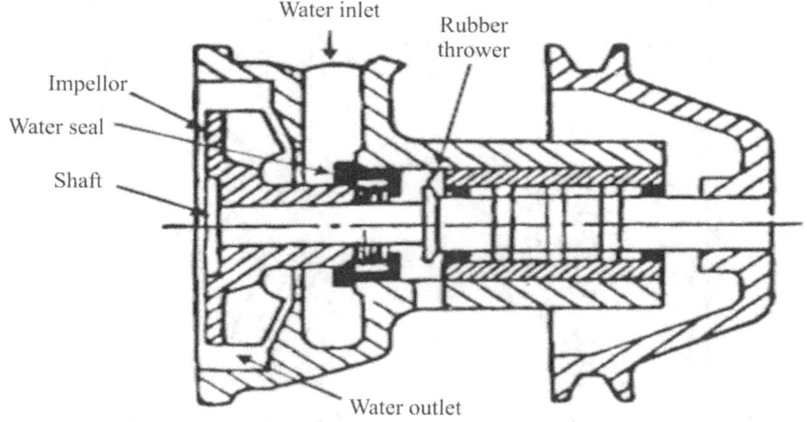

Fig. 2.15 Water Pump

Fan: It is driven by the engine output shaft through same belt that drives the pump. It is provided behind the radiator and it blows air over the radiator for cooling purpose.

Water Jackets: Cooling water jackets are provided around the cylinder, cylinder head, and valve seat and any hot parts which are to be cooled. Heat generated in the engine cylinder, conducted through the cylinder walls to the jackets. The water flowing through the jackets absorbs this heat and gets hot. This hot water will then be cooled in the radiator (Referred Figure 2.16).

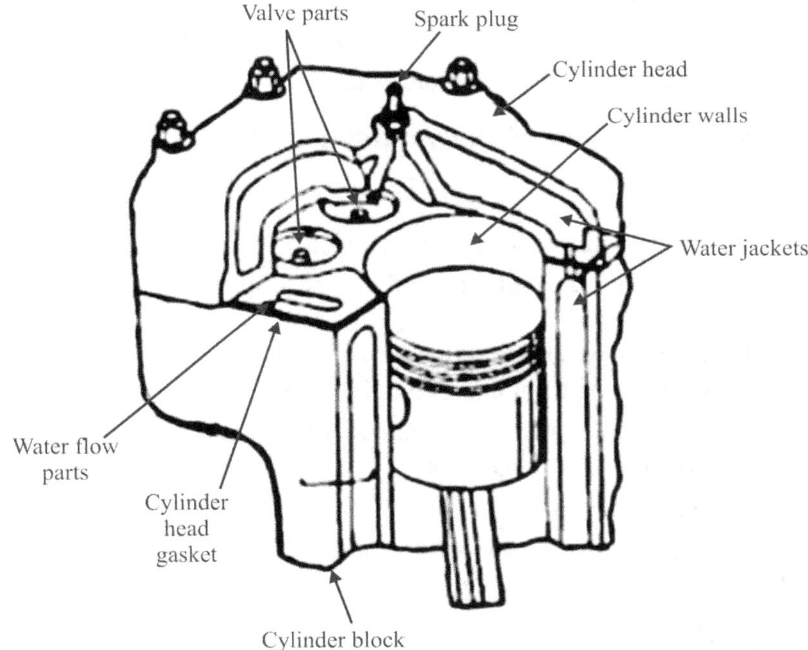

Fig. 2.16 Water Jackets

Antifreeze Mixture: In western countries if the water used in the radiator freezes because of cold climates, then ice formed has more volume and produces cracks in the cylinder blocks, pipes, and radiator. So, to prevent freezing antifreeze mixtures or solutions are added in the cooling water.

The ideal antifreeze solutions should have the following properties :

    (a)   It should dissolve in water easily.

    (b)   It should not evaporate.

    (c)   It should not deposit any foreign matter in cooling system.

(d) It should not have any harmful effect on any part of cooling system.

(e) It should be cheap and easily available.

(f) It should not corrode the system.

No single antifreeze satisfies all the requirements. Normally following are used as antifreeze solutions:

(a) Methyl, ethyl and isopropyl alcohols.

(b) A solution of alcohol and water.

(c) Ethylene Glycol.

(d) A solution of water and Ethylene Glycol.

(e) Glycerine along with water, etc.

2.2.4 Advantages and Disadvantages of Water Cooling Systems

Advantages

(a) Uniform cooling of cylinder, cylinder head and valves.

(b) Specific fuel consumption of engine improves by using water cooling system.

(c) If we employ water cooling system, then engine need not be provided at the front end of moving vehicle.

(d) Engine is less noisy as compared with air cooled engines, as it has water for damping noise.

Disadvantages

(a) It depends upon the supply of water.

(b) The water pump which circulates water absorbs considerable power.

(c) If the water cooling system fails then it will result in severe damage of engine.

(d) The water cooling system is costlier as it has more number of parts. Also it requires more maintenance and care for its parts.

Review Questions

1. Differentiate between battery ignition and magneto ignition system.

2. List the various types of ignition system.

3. Discuss the basic requirement of spark ignition system.

4. What is mean by firing order in an IC engine?

5. Discuss the limitation of conventional ignition system.

6. List the various types of engine cooling system.

7. Discuss the working principal of evaporative and pressure cooling system.

8. List the advantages and disadvantages of liquid and air cooling system.

9. List the various anti freeze solution used in cooling system.

CHAPTER 3

IC Engine Testing and Performance

3.1 Introduction

At a design and development stage an engineer would design an engine with certain aims in his mind. The aims may include the variables like indicated power, brake power, brake specific fuel consumption, exhaust emissions, cooling of engine, maintenance free operation etc. The other task of the development engineer is to reduce the cost and improve power output and reliability of an engine. In trying to achieve these goals he has to try various design concepts. After the design, the parts of the engine are manufactured for the dimensions and surface finish and may be with certain tolerances. In order verify the designed and developed engine one has to go for testing and performance evaluation of the engines.

Thus, in general, a development engineer will have to conduct a wide variety of engine tests starting from simple fuel and air-flow measurements to taking of complicated injector needle lift diagrams, swirl patterns and photographs of the burning process in the combustion chamber. The nature and the type of the tests to be conducted depend upon various factors, some of which are: the degree of development of the particular design, the accuracy required, the funds available, the nature of the manufacturing company, and its design strategy. In this chapter, only certain basic tests and measurements will be considered.

3.2   Performance Parameters

Engine performance is an indication of the degree of success of the engine performs its assigned task, i.e., the conversion of the chemical energy contained in the fuel into the useful mechanical work. The performance of an engine is evaluated on the basis of the following:

    (a)   Specific Fuel Consumption.

    (b)   Brake Mean Effective Pressure.

    (c)   Specific Power Output.

    (d)   Specific Weight.

    (e)   Exhaust Smoke and other Emissions.

The particular application of the engine decides the relative importance of these performance parameters. For Example: for an aircraft engine specific weight is more important whereas for an industrial engine specific fuel consumption is more important.

For the evaluation of an engine performance few more parameters are chosen and the effects of various operating conditions, design concepts and modifications on these parameters are studied. The basic performance parameters are the following:

    (a)   Power and Mechanical Efficiency.

    (b)   Mean Effective Pressure and Torque.

    (c)   Specific Output.

    (d)   Volumetric Efficiency.

    (e)   Fuel-air Ratio.

    (f)   Specific Fuel Consumption.

    (g)   Thermal Efficiency and Heat Balance.

    (h)   Exhaust Smoke and other Emissions.

    (i)   Specific Weight.

Power and Mechanical Efficiency

The main purpose of running an engine is to obtain mechanical power.

- Power is defined as the rate of doing work and is equal to the product of force and linear velocity or the product of torque and angular velocity.

- Thus, the measurement of power involves the measurement of force (or torque) as well as speed. The force or torque is measured with the help of a dynamometer and the speed by a tachometer. The power developed by an engine and measured at the output shaft is called the brake power (BP) and is given by

$$BP = \frac{2\Pi NT}{60} \quad\quad .....(3.1)$$

where

T = torque (N – m) and

N = the rotational speed (rpm)

The total power developed by combustion of fuel in the combustion chamber is, however, more than the BP and is called indicated power (IP). of the power developed by the engine, i.e., IP, some power is consumed in overcoming the friction between moving parts, some in the process of inducting the air and removing the products of combustion from the engine combustion chamber.

Indicated Power IC Engine Testing

It is the power developed in the cylinder and thus, forms the basis of evaluation of combustion efficiency or the heat release in the cylinder.

$$IP = \frac{P_{im}LANk}{60}$$

where,

P_{im} = Mean effective pressure, N/m^2,

L = Length of the stroke, m,

A = Area of the piston, m^2,

N = Rotational speed of the engine, rpm (It is N/2 for four stroke engine), and

k = Number of cylinders.

Thus, we see that for a given engine the power output can be measured in terms of mean effective pressure. The difference between the ip and bp is the indication of the power lost in the mechanical components of the engine (due to friction) and forms the basis of mechanical efficiency; which is defined as follows:

$$Mechanical\ Efficiency = \frac{BP}{IP} \quad\quad .....(3.2)$$

The difference between IP and BP is called friction power (FP).

$$FP = IP - BP \qquad \qquad .....(3.3)$$

$$\therefore \quad \text{Mechanical efficiency} \quad = \frac{BP}{BP + FP} \qquad \qquad .....(3.4)$$

Mean Effective Pressure and Torque

Mean effective pressure is defined as a hypothetical/average pressure which is assumed to be acting on the piston throughout the power stroke. Therefore,

$$P_{im} = \frac{IP \times 60}{LANk} \qquad \qquad .....(3.5)$$

where,

P_m = Mean effective pressure, N/m^2,

IP = Indicated power, Watt,

L = Length of the stroke, m,

A = Area of the piston, m^2,

N = Rotational speed of the engine, rpm (It is N/2 for four stroke engine), and

k = Number of cylinders.

If the mean effective pressure is based on BP it is called the brake mean effective pressure (B_{mep}), and if based on IP it is called indicated mean effective pressure (I_{mep}). Similarly, the friction mean effective pressure (F_{mep}) can be defined as,

$$F_{mep} = I_{mep} - B_{mep} \qquad \qquad .....(3.6)$$

The torque is related to mean effective pressure by the relation

$$BP = \frac{2\pi NT}{60}$$

$$IP = \frac{P_{im} LANk}{60} \qquad \qquad .....(3.7)$$

By Eq. (3.5),

$$\frac{2\pi NT}{60} = B_{mep} \times A \times L \times \frac{Nk}{60}$$

$$T = \frac{B_{mep} \times A \times L \times k}{2\pi} \qquad \qquad .....(3.8)$$

Thus, the torque and the mean effective pressure are related by the engine size. A large engine produces more torque for the same mean effective pressure. For this reason, torque is not the measure of the ability of an engine to utilize its displacement for producing power from fuel. It is the mean effective pressure which gives an indication of engine displacement utilization for this conversion. Higher the mean effective pressure, higher will be the power developed by the engine for a given displacement.

Again we see that the power of an engine is dependent on its size and speed. Therefore, it is not possible to compare engines on the basis of either power or torque. Mean effective pressure is the true indication of the relative performance of different engines.

Specific Output

Specific output of an engine is defined as the brake power (output) per unit of piston displacement and is given by,

$$\text{Specific Output} = \frac{BP}{A \times L}$$

$$= \text{Constant} \times B_{mep} \times \text{rpm} \quad\quad\quad .....(3.9)$$

- The specific output consists of two elements – the B_{mep} (force) available to work and the speed with which it is working.

- Therefore, for the same piston displacement and B_{mep} an engine operating at higher speed will give more output.

- It is clear that the output of an engine can be increased by increasing either speed or B_{mep}. Increasing speed involves increase in the mechanical stress of various engine parts whereas increasing B_{mep} requires better heat release and more load on engine cylinder.

Volumetric Efficiency (η_v)

Volumetric efficiency of an engine is an indication of the measure of the degree to which the engine fills its swept volume. It is defined as the ratio of the mass of air inducted into the engine cylinder during the suction stroke to the mass of the air corresponding to the swept volume of the engine at atmospheric pressure and temperature. Alternatively, it can be defined as the ratio of the actual volume inhaled during suction stroke measured at intake conditions to the swept volume of the piston.

$$\eta_v = \frac{\text{Mass of charge ctually sucked in}}{\text{Mass of ch arg e corresponding to the cylinder intake P and T conditions}} \quad .....(3.10)$$

The amount of air taken inside the cylinder is dependent on the volumetric efficiency of an engine and hence puts a limit on the amount of fuel which can be efficiently burned and the power output. For supercharged engine the volumetric efficiency has no meaning as it comes out to be more than unity.

Fuel-Air Ratio (F/A)

Fuel-air ratio (F/A) is the ratio of the mass of fuel to the mass of air in the fuel-air mixture. Air-fuel ratio (A/F) is reciprocal of fuel-air ratio. Fuel-air ratio of the mixture affects the combustion phenomenon in that it determines the flame propagation velocity, the heat release in the combustion chamber, the maximum temperature and the completeness of combustion. Relative fuel-air ratio is defined as the ratio of the actual fuel-air ratio to that of the stoichiometric fuel-air ratio required to burn the fuel supplied. Stoichiometric fuel-air ratio is the ratio of fuel to air is one in which case fuel is completely burned due to minimum quantity of air supplied.

$$\text{Relative F/A ratio, FR} = \frac{\text{Actual F/A Ratio}}{\text{Stoichometric F/A Ratio}} \quad \text{.....(3.11)}$$

Brake Specific Fuel Consumption

Brake specific fuel consumption is defined as the amount of fuel consumed for each unit of brake power developed per hour. It is a clear indication of the efficiency with which the engine develops power from fuel.

$$\text{Brake specific fuel consumption (bsfc)} = \frac{\text{Fuel Consumption}}{\text{BP}} \quad \text{.....(3.12)}$$

This parameter is widely used to compare the performance of different engines. Thermal Efficiency and Heat Balance Thermal efficiency of an engine is defined as the ratio of the output to that of the chemical energy input in the form of fuel supply. It may be based on brake or indicated output. It is the true indication of the efficiency with which the chemical energy of fuel (input) is converted into mechanical work. Thermal efficiency also accounts for combustion efficiency, i.e., for the fact that whole of the chemical energy of the fuel is not converted into heat energy during combustion.

$$\text{Brake thermal efficiency} = \frac{\text{BP}}{m_f \times C_v} \quad \text{.....(3.13)}$$

where,

C_v = Calorific value of fuel, kJ/kg, and

m_f = Mass of fuel supplied, kg/sec.

- The energy input to the engine goes out in various forms – a part is in the form of brake output, a part into exhaust, and the rest is taken by cooling water and the lubricating oil.
- The break-up of the total energy input into these different parts is called the heat balance.
- The main components in a heat balance are brake output, coolant losses, heat going to exhaust, radiation and other losses.
- Preparation of heat balance sheet gives us an idea about the amount of energy wasted in various parts and allows us to think of methods to reduce the losses so incurred.

Exhaust Smoke and other Emissions

Smoke and other exhaust emissions such as oxides of nitrogen, unburned hydrocarbons, etc., are nuisance for the public environment. With increasing emphasis on air pollution control, all efforts are being made to keep them as minimum as it could be. Smoke is an indication of incomplete combustion. It limits the output of an engine if air pollution control is the consideration.

Exhaust emissions have of late become a matter of grave concern and with the enforcement of legislation on air pollution in many countries; it has become necessary to view them as performance parameters.

Specific Weight

Specific weight is defined as the weight of the engine in kilogram for each brake power developed and is an indication of the engine bulk. Specific weight plays an important role in applications such as power plants for aircrafts.

3.3 Basic Measurements

The basic measurements to be undertaken to evaluate the performance of an engine on almost all tests are the following:

(a)   Speed

(b)   Fuel consumption

(c)   Air consumption

(d)   Smoke density

(e) Brake horse-power

(f) Indicated horse power and friction horse power

(g) Heat going to cooling water

(h) Heat going to exhaust

(i) Exhaust gas analysis.

In addition to above a large number of other measurements may be necessary depending upon the aim of the test.

3.3.1 Measurement of Speed

One of the basic measurements is that of speed. A wide variety of speed measuring devices are available in the market. They range from a mechanical tachometer to digital and triggered electrical tachometers. The best method of measuring speed is to count the number of revolutions in a given time. This gives an accurate measurement of speed. Many engines are fitted with such revolution counters.

A mechanical tachometer or an electrical tachometer can also be used for measuring the speed. The electrical tachometer has a three-phase permanent-magnet alternator to which a voltmeter is attached. The output of the alternator is a linear function of the speed and is directly indicated on the voltmeter dial.

Both electrical and mechanical types of tachometers are affected by the temperature variations and are not very accurate. For accurate and continuous measurement of speed a magnetic pick-up placed near a toothed wheel coupled to the engine shaft can be used. The magnetic pick-up will produce a pulse for every revolution and a pulse counter will accurately measure the speed.

3.3.2 Fuel Consumption Measurement

Fuel consumption is measured in two ways:

(a) The fuel consumption of an engine is measured by determining the volume flow in a given time interval and multiplying it by the specific gravity of the fuel which should be measured occasionally to get an accurate value.

(b) Another method is to measure the time required for consumption of a given mass of fuel.

Accurate measurement of fuel consumption is very important in engine testing work.

As already mentioned two basic types of fuel measurement methods are:

- Volumetric type
- Gravimetric type.

Volumetric type flow-meter includes Burette method, Automatic Burette flowmeter and turbine flowmeter.

Gravimetric Fuel Flow Measurement: The efficiency of an engine is related to the kilograms of fuel which are consumed and not the number of litres. The method of measuring volume flow and then correcting it for specific gravity variations is quite inconvenient and inherently limited in accuracy. Instead if the weight of the fuel consumed is directly measured a great improvement in accuracy and cost can be obtained. There are three types of gravimetric type systems which are commercially available include Actual weighing of fuel consumed, Four Orifice Flowmeter, etc.

3.3.3 Measurement of Air Consumption

One can say the mixture of air and fuel is the food for an engine. For finding out the performance of the engine, accurate measurement of both is essential. In IC engines, the satisfactory measurement of air consumption is quite difficult because the flow is pulsating, due to the cyclic nature of the engine and because the air a compressible fluid. Therefore, the simple method of using an orifice in the induction pipe is not satisfactory since the reading will be pulsating and unreliable. All kinetic flow-inferring systems such as nozzles, orifices and venturies have a square law relationship between flow rate and differential pressure which gives rise to severe errors on unsteady flow. Pulsation produced errors are roughly inversely proportional to the pressure across the orifice for a given set of flow conditions. The various methods and meters used for air flow measurement include

(a)   Air box method, and

(b)   Viscous-flow air meter.

3.3.4 Measurement of Exhaust Smoke

All the three widely used smokemeters, namely, Bosch, Hartridge, and PHS are basically soot density (g/m^3) measuring devices, that is, the meter readings are a function of the mass of carbon in a given volume of exhaust gas. Hartridge smokemeter works on the light extinction principle.

The basic principles of the Bosch smokemeter is one in which a fixed quantity of exhaust gas is passed through a fixed filter paper and the density of the smoke stains on the paper are evaluated optically. In a recent modification of this type of smokemeter units are used for the measurement of the intensity of smoke stain on filter paper. In Von Brand smokemeter which can give a continuous reading a filter tape is continuously moved at a uniform rate to which the exhaust from the engine is fed. The smoke stains developed on the filter paper are sensed by a recording head. The signal obtained from the recording head is calibrated to give smoke density.

3.4 Measurement of Exhaust Emission

Substances which are emitted to the atmosphere from any opening of the exhaust port of the engine are termed as exhaust emissions. If combustion is complete and the mixture is stoichiometric the products of combustion would consist of carbon dioxide (CO_2) and water vapour only. However, there is no complete combustion of fuel and hence the exhaust gas consists of variety of components, the most important of them are carbon monoxide (CO), unburned hydrocarbons (UBHC) and oxides of nitrogen (NOx). Some oxygen and other inert gases would also be present in the exhaust gas.

3.5 Measurement of Brake Power

The brake power measurement involves the determination of the torque and the angular speed of the engine output shaft. The torque measuring device is called a dynamometer. Dynamometers can be broadly classified into two main types, power absorption dynamometers and transmission dynamometer.

Fig. 3.1 shows the basic principle of a dynamometer. A rotor driven by the engine under test is electrically, hydraulically or magnetically coupled to a stator. For every revolution of the shaft, the rotor periphery moves through a distance 2πR against the coupling force F. Hence, the work done per revolution is:

$$W = 2\,\pi RF$$

The external moment or torque is equal to $S \times L$ where, S is the scale reading and L is the arm. This moment balances the turning moment $R \times F$, i.e.,

$$S \times L = R \times F$$

$$\text{Work done/revolution} = 2\pi\,SL$$

$$\therefore \quad \text{Work done/minute} = 2\pi\,SLN$$

where, N is rpm. Hence, power is given by

Brake power $P = 2 \pi NT$

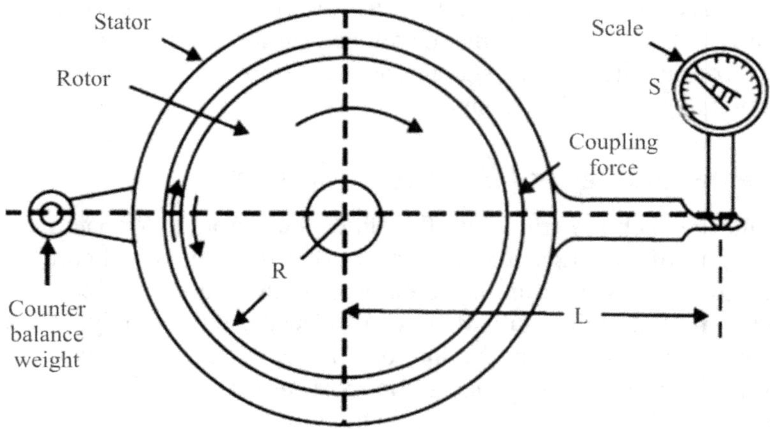

Fig. 3.1 Principle of a Dynamometer

Absorption Dynamometers

These dynamometers measure and absorb the power output of the engine to which they are coupled. The power absorbed is usually dissipated as heat by some means. Example of such dynamometers is prony brake, rope brake, hydraulic dynamometer, etc.

Transmission Dynamometers

In transmission dynamometers, the power is transmitted to the load coupled to the engine after it is indicated on some type of scale. These are also called torque-meters.

3.5.1 Absorption Dynamometers

These include Prony brake type, Rope brake type, and Hydraulic type.

Prony Brake

One of the simplest methods of measuring brake power (output) is to attempt to stop the engine by means of a brake on the flywheel and measure the weight which an arm attached to the brake will support, as it tries to rotate with the flywheel. This system is known as the prony brake and forms its use; the expression brake power has come.

The Prony brake is shown in Fig. 3.2. This works on the principle of converting power into heat by dry friction. It consists of wooden block mounted on a flexible rope or band the wooden block when pressed into contact with the rotating drum takes the engine torque and the power is dissipated in frictional resistance. Spring-loaded bolts are provided to tighten the wooden block and hence increase the friction. The whole of the power absorbed is converted into heat and hence this type of dynamometer must the cooled. The brake horsepower is given by

$$BP = 2\,\pi\,NT$$

where $T = W \times l$

W being the weight applied at a radius l.

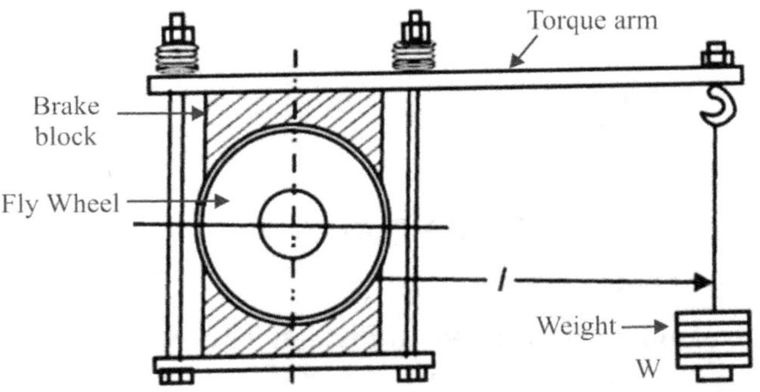

Fig. 3.2 Pony Brake

Rope Brake

The rope brake as shown in Fig. 3.3 is another simple device for measuring bp of an engine. It consists of a number of turns of rope wound around the rotating drum attached to the output shaft. One side of the rope is connected to a spring balance and the other to a loading device. The power is absorbed in friction between the rope and the drum. The drum therefore requires cooling.

Rope brake is cheap and easily constructed but not a very accurate method because of changes in the friction coefficient of the rope with temperature.

The BP is given by

$$BP = \pi\,DN\,(W - S)$$

where, D is the brake drum diameter, W is the weight in Newton and S is the spring scale reading.

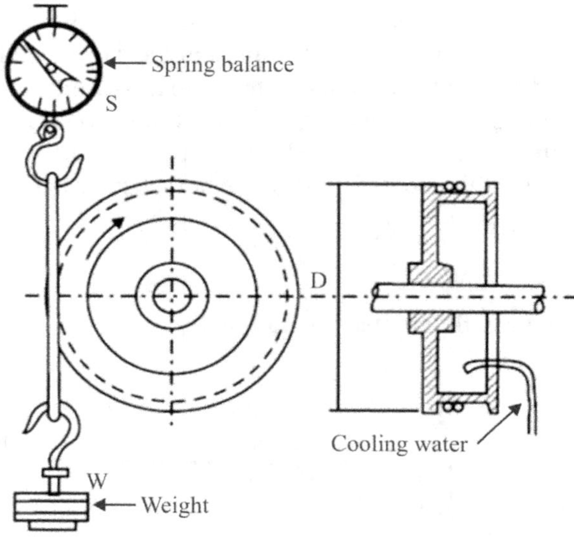

Fig. 3.3 Rope Brake

Hydraulic Dynamometer

Hydraulic dynamometer is shown in Fig.3.4. This works on the principle of dissipating the power in fluid friction rather than in dry friction.

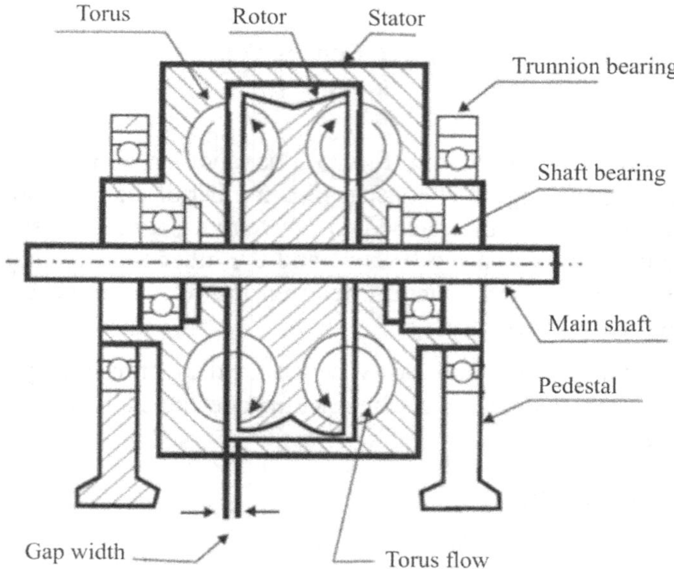

Fig. 3.4 Hydraulic Dynamometer

- In principle its construction is similar to that of a fluid flywheel.
- It consists of an inner rotating member or impeller coupled to the output shaft of the engine.
- This impeller rotates in a casing filled with fluid.
- This outer casing, due to the centrifugal force developed, tends to revolve with the impeller, but is resisted by a torque arm supporting the balance weight.
- The frictional forces between the impeller and the fluid are measured by the spring-balance fitted on the casing.
- The heat developed due to dissipation of power is carried away by a continuous supply of the working fluid, usually water.
- The output can be controlled by regulating the sluice gates which can be moved in and out to partially or wholly obstruct the flow of water between impeller, and the casing.

Eddy Current Dynamometer

The working principle of eddy current dynamometer is shown in Figure 3.5. It consists of a stator on which are fitted a number of electromagnets and a rotor disc made of copper or steel and coupled to the output shaft of the engine. When the rotor rotates eddy currents are produced in the stator due to magnetic flux set up by the passage of field current in the electromagnets. These eddy currents are dissipated in producing heat so that this type of dynamometer also requires some cooling arrangement. The torque is measured exactly as in other types of absorption dynamometers, i.e., with the help of a moment arm. The load is controlled by regulating the current in the electromagnets.

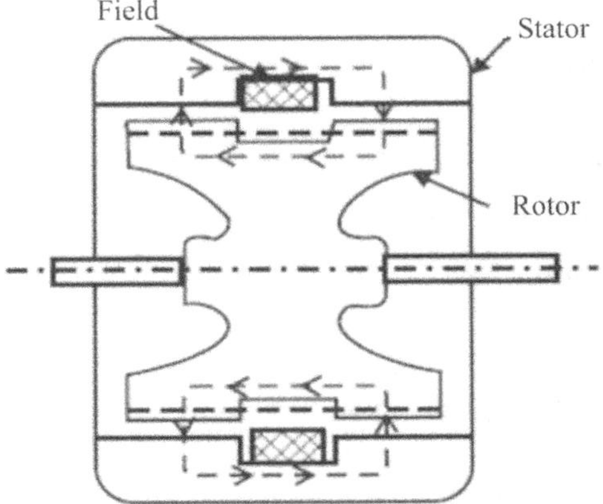

Fig. 3.5 Eddy Current Dynamometer

The following are the main advantages of eddy current dynamometers:

(a)   High brake power per unit weight of dynamometer.

(b)   They offer the highest ratio of constant power speed range (up to 5:1).

(c)   Level of field excitation is below 1% of total power being handled by dynamometer, thus, easy to control and program.

(d)   Development of eddy current is smooth hence the torque is also smooth and continuous under all conditions.

(e)   Relatively higher torque under low speed conditions.

(f)   It has no intricate rotating parts except shaft bearing.

(g)   No natural limit to size – either small or large.

Swinging Field D.C. Dynamometer

Basically, a swinging field d.c. dynamometer is a d.c. shunt motor so supported on trunnion bearings to measure their action torque that the outer case and filed coils tend to rotate with the magnetic drag. Hence, the name swinging field. The torque is measured with an arm and weighing equipment in the usual manner.

Many dynamometers are provided with suitable electric connections to run as motor also. Then the dynamometer is reversible, i.e. works as motoring as well as power absorbing device.

- When used as an absorption dynamometer it works as a d.c. generator and converts mechanical energy into electric energy which is dissipated in an external resistor or fed back to the mains.

- When used as a motoring device an external source of d.c. voltage is needed to drive the motor.

The load is controlled by changing the field current.

3.5.2 Fan Dynamometer

It is also an absorption type of dynamometer in that when driven by the engine it absorbs the engine power. Such dynamometers are useful mainly for rough testing and running in. The accuracy of the fan dynamometer is very poor. The power absorbed is determined by using previous calibration of the fan brake.

3.5.3 Transmission Dynamometers

Transmission dynamometers, also called torque meters, mostly consist of a set of strain-gauges fixed on the rotating shaft and the torque is measured by the angular deformation of the shaft which is indicated as strain of the strain gauge. Usually, a four arm bridge is used to reduce the effect of temperature to minimum and the gauges are arranged in pairs such that the effect of axial or transverse load on the strain gauges is avoided.

Fig. 3.6 shows a transmission dynamometer which employs beams and strain-gauges for a sensing torque. Transmission dynamometers are very accurate and are used where continuous transmission of load is necessary. These are used mainly in automatic units.

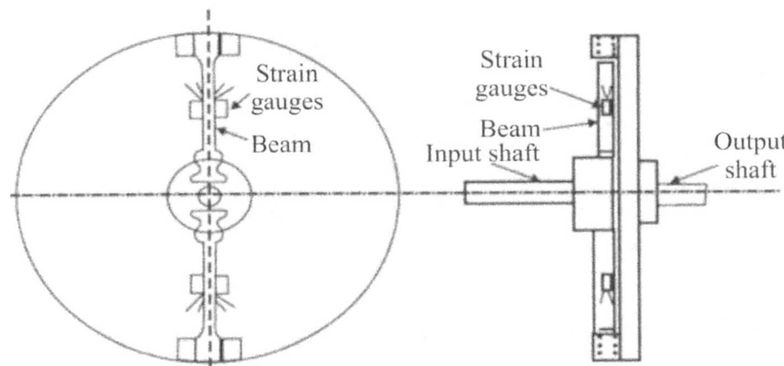

Fig. 3.6 Transmission Dynamometer

3.6 Measurement of Friction Horse Power

- The difference between indicated power and the brake power output of an engine is the friction power.

- Almost invariably, the difference between a good engine and a bad engine is due to difference between their frictional losses.

- The frictional losses are ultimately dissipated to the cooling system (and exhaust) as they appear in the form of frictional heat and this influences the cooling capacity required. Moreover, lower friction means availability of more brake power; hence brake specific fuel consumption is lower.

- The bsfc rises with an increase in speed and at some speed it decreases. Thus, the level of friction decides the maximum output of the engine which can be obtained economically.

In the design and testing of an engine; measurement of friction power is important for getting an insight into the methods by which the output of an engine can be increased. In the evaluation of IP and mechanical efficiency measured friction power is also used.

The friction force power of an engine is determined by the following methods:

(a) Willian's line method.

(b) Morse test.

(c) Motoring test.

(d) Difference between IP and BP.

Willan's Line Method or Fuel Rate Extrapolation

In this method, gross fuel consumption versus BP at a constant speed is plotted and the graph is extrapolated back to zero fuel consumption as illustrated in Fig. 3.7. At the point where this graph cuts the BP axis is an indication of the friction power of the engine at that speed. This negative work represents the combined loss due to mechanical friction, pumping and blow-by.

The test is applicable only to compression ignition engines.

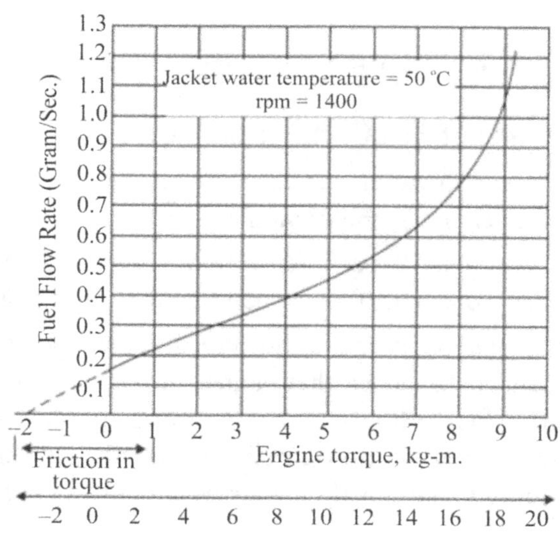

Fig. 3.7 Willan's Line Method

- The main drawback of this method is the long distance to be extrapolated from data measured between 5 and 40% load towards the zero line of fuel input.

- The directional margin of error is rather wide because of the graph which may not be a straight line many times.

- The changing slope along the curve indicates part efficiencies of increments of fuel. The pronounced change in the slope of this line near full load reflects the limiting influence of the air-fuel ratio and of the quality of combustion.

- Similarly, there is a slight curvature at light loads. This is perhaps due to difficulty in injecting accurately and consistently very small quantities of fuel per cycle.

- Therefore, it is essential that great care should be taken at light loads to establish the true nature of the curve.

- The Willan's line for a swirl-chamber CI engine is straighter than that for a direct injection type engine.

- The accuracy obtained in this method is good and compares favorably with other methods if extrapolation is carefully done.

Morse Test

The Morse test is applicable only to multi-cylinder engines.

- In this test, the engine is first run at the required speed and the output is measured.

- Then, one cylinder is cut out by short circuiting the spark plug or by disconnecting the injector as the case may be.

- Under this condition all other cylinders 'motor' this cut-out cylinder. The output is measured by keeping the speed constant at its original value.

- The difference in the outputs is a measure of the indicated horse power of the cut-out cylinder.

- Thus, for each cylinder the IP is obtained and is added together to find the total IP of the engine.

The IP of n cylinder is given by

$$IP_n = BP_n + FP \qquad \qquad .....(3.14)$$

IP for (n − 1) cylinders is given by

$$IP_{n-1} = BP_{n-1} + FP \qquad \qquad .....(3.15)$$

Since, the engine is running at the same speed it is quite reasonable to assume that FP remains constant.

From Eqs. (3.17) and (3.18), we see that the IP of the nth cylinder is given by

$$(IP)n^{th} = BP_n - BP_{n-1} \qquad .....(3.16)$$

and the total IP of the engine is,

$$IP_n = \sum_{i=1}^{n} IPn^{th} \qquad .....(3.17)$$

By subtracting BP_n from this, FP of the engine can be obtained. This method though gives reasonably accurate results and is liable to errors due to changes in mixture distribution and other conditions by cutting-out one cylinder.

In gasoline engines, where there is a common manifold for two or more cylinders the mixture distribution as well as the volumetric efficiency both change. Again, almost all engines have a common exhaust manifold for all cylinders and cutting-out of one cylinder may greatly affect the pulsations in exhaust system which may significantly change the engine performance by imposing different back pressures.

Motoring Test

- In the motoring test, the engine is first run up to the desired speed by its own power and allowed to remain at the given speed and load conditions for some time so that oil, water, and engine component temperatures reach stable conditions.

- The power of the engine during this period is absorbed by a swinging field type electric dynamometer, which is most suitable for this test.

- The fuel supply is then cut-off and by suitable electric-switching devices the dynamometer is converted to run as a motor to drive for 'motor' the engine at the same speed at which it was previously running.

- The power supply to the motor is measured which is a measure of the fhp of the engine. During the motoring test the water supply is also cut-off so that the actual operating temperatures are maintained.

- This method, though determines the fp at temperature conditions very near to the actual operating temperatures at the test speed and load, does, not give the true losses occurring under firing conditions due to the following reasons.

  (a) The temperatures in the motored engine are different from those in a firing engine because even if water circulation is stopped the incoming air cools the cylinder. This reduces the lubricating oil temperature and increases friction increasing the oil viscosity. This problem is much more sever in air-cooled engines.

(b)   The pressure on the bearings and piston rings is lower than the firing pressure. Load on main and connecting road bearings are lower.

(c)   The clearance between piston and cylinder wall is more (due to cooling). This reduces the piston friction.

(d)   The air is drawn at a temperature less than when the engine is firing because it does not get heat from the cylinder (rather loses heat to the cylinder). This makes the expansion line to be lower than the compression line on the P-V diagram. This loss is however counted in the indicator diagram.

(e)   During exhaust the back pressure is more because under motoring IC Engine Testing conditions sufficient pressure difference is not available to impart gases; the kinetic energy is necessary to expel them from exhaust.

Motoring method, however, gives reasonably good results and is very suitable for finding the losses due to various engine components. This insight into the losses caused by various components and other parameters is obtained by progressive stripping-off of the under progressive dismantling conditions keeping water and oil circulation intact. Then the cylinder head can be removed to evaluate, by difference, the compression loss. In this manner piston ring, piston etc. can be removed and evaluated for their effect on overall friction.

Difference between IP and BP

(a)   The method of finding the FP by computing the difference between IP, as obtained from an indicator diagram, and BP, as obtained by a dynamometer, is the ideal method.

(b)   In obtaining accurate indicator diagrams, especially at high engine speeds, this method is usually only used in research laboratories. Its use at commercial level is very limited.

Comments on Methods of Measuring FP

- The Willan's line method and Morse tests are very cheap and easy to conduct.

- However, both these tests give only an overall idea of the losses whereas motoring test gives a very good insight into the various causes of losses and is a much more powerful tool.

- As far as accuracy is concerned the IP–BP method is the most accurate if carefully done.

- Motoring method usually gives a higher value for FP as compared to that given by the Willian's line method.

3.7 Blow-by Loss

Blow-by is the escape of unburned air-fuel mixture and burned gases from the combustion chamber, past the piston rings, and into the crankcase. High blow-by is quite harmful in that it results in higher ring temperatures and contamination of lubricating oil.

3.8 Performance of SI Engines

The performance of an engine is usually studied by heat balance-sheet. The main components of the heat balance are:

- Heat equivalent to the effective (brake) work of the engine,
- Heat rejected to the cooling medium,
- Heat carried away from the engine with the exhaust gases, and
- Unaccounted losses.

The unaccounted losses include the radiation losses from the various parts of the engine and heat lost due to incomplete combustion. The friction loss is not shown as a separate item to the heat balance-sheet as the friction loss ultimately reappears as heat in cooling water, exhaust and radiation.

The Table 3.1 gives the approximate percentage values of various losses in SI and CI engines.

Table 3.1

Engine Type	Brake load efficiency %	Heat rejected to cooling water %	Heat rejected through exhaust gases %	Unaccounted heat %
SI	21-28	12-27	30-55	3-55 (including incomplete combustion loss 0-45)
CI	29-42	15-35	25-45	21-0 (including incomplete combustion loss 0-5)

Figure 3.8 shows the heat balance for a petrol engine run at full throttle over its speed range. In SI engines, the loss due to incomplete combustion included on unaccounted form can be rather high. For a rich mixture (A/F ratio = 12.5 to 13) it could be 20%.

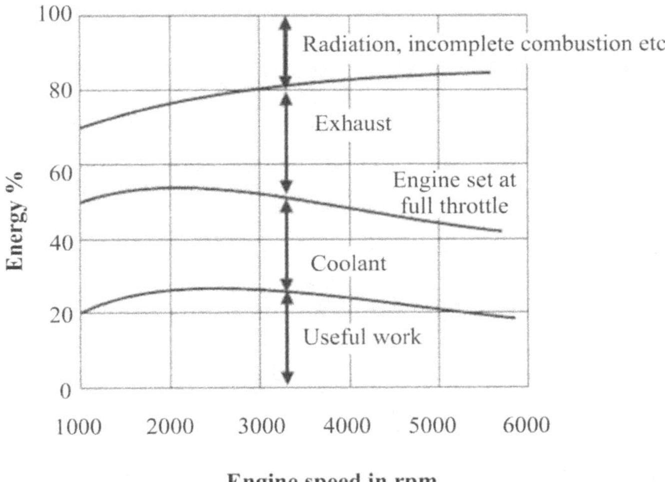

Engine speed in rpm

Fig. 3.8 Heat Balance Vs Speed for a Petrol Engine at Full Throttle

Fig. 3.9 shows the heat balance of uncontrolled Otto engine at different loads.

Fig. 3.10 shows the brake thermal efficiency, indicated thermal efficiency, mechanical efficiency and specific fuel consumption for the above SI engine.

Fig. 3.11 shows the IP, BP, FP (by difference) brake torque, brake mean effective pressure and brake specific fuel consumption of a high compression ratio (9) automotive SI engine at full or Wide Open Throttle (W.O.T.).

Speed for a Petrol Engine at Full Throttle

Referring to the Fig. 3.7 through Fig. 3.11 the following conclusions can be drawn:

(a) At full throttle the brake thermal efficiency at various speeds varies from 20 to 27 percent, maximum efficiency being at the middle speed range.

(b) The percentage heat rejected to coolant is more at lower speed (>> 35 percent) and reduces at higher speeds (>> 25 percent). Considerably more heat is carried by exhaust at higher speeds.

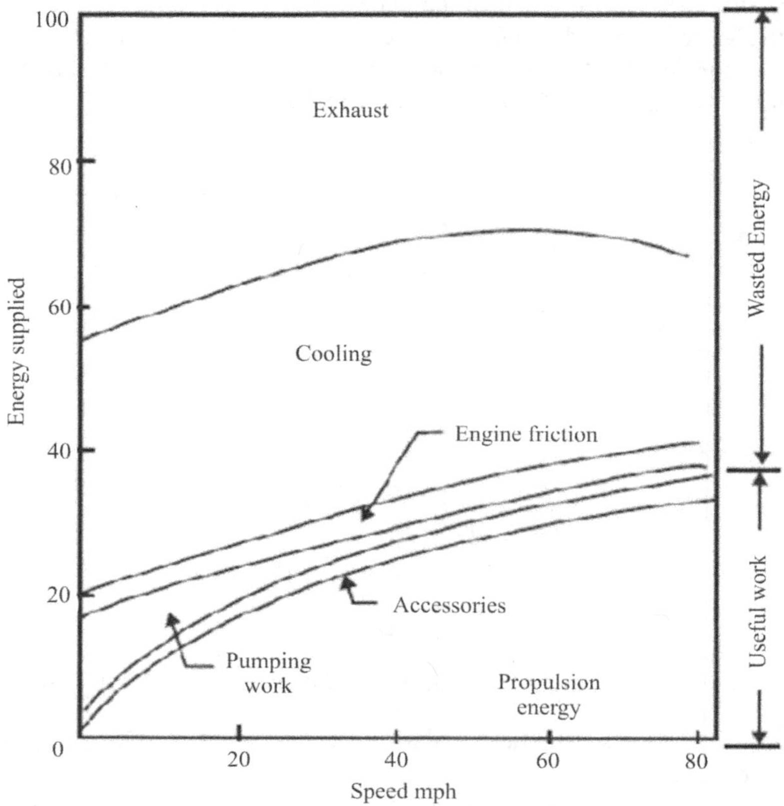

Fig. 3.9 Uncontrolled Otto Engine

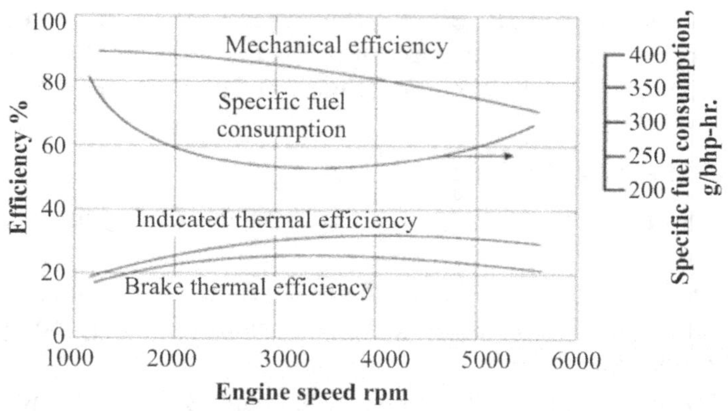

Fig. 3.10 Efficiency and Specific Fuel Consumption

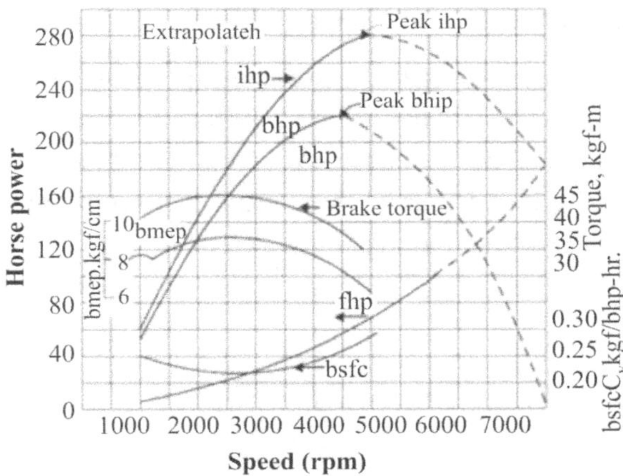

Fig. 3.11 Variable Speed Test of Automotive SI Engine at Full Throttle (CR = 9)

(c) Torque and mean effective pressure do not strongly depend on the speed of the engine, but depend on the volumetric efficiency and friction losses. Maximum torque position corresponds with the maximum air charge or minimum volumetric efficiency position. Torque and mean effective pressure curves peak at about half that of the brake-power.

Note: If size (displacement) of the engine were to be doubled, torque would also double, but mean effective pressure (mep) is a 'specific' torque, a variable independent of the size of the engine.

(d) High power arises from the high speed. In the speed range before the maximum power is obtained, doubling the speed doubles the power.

(e) At low engine speed the friction power is relatively low and BP is nearly as large as IP (Fig. 3.13). As engine speed increases, however, FP increases at continuously greater rate and therefore BP reaches a peak and starts reducing even though IP is rising. At engine speeds above the usual operating range, FP increases very rapidly. Also, at these higher speeds IP will reach a maximum and then fall off. At some point, IP and FP will be equal, and BP will then drop to zero.

Performance of SI Engine at Constant Speed and Variable Load

The performance of SI engine at constant speed and variable loads is different from the performance at full throttle and variable speed. Fig. 3.12 shows the heat balance of SI engine at constant speed and variable load. The load is varied by altering the throttle and the speed is kept constant by resetting the dynamometer.

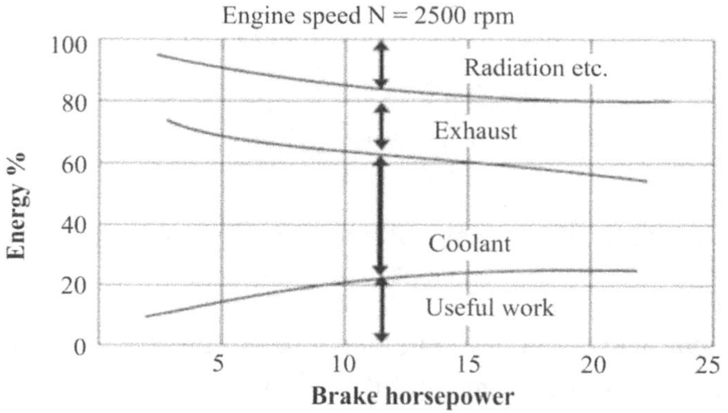

Fig. 3.12 Heat Balance Vs Load for a Petrol Engine

Closing the throttle reduces the pressure inside the cylinders but the temperature is affected very little because the air/fuel ratio is substantially constant, and the gas temperatures throughout the cycle are high. This results in high loss to coolant at low engine load. This is reason of poor part load thermal efficiency of the SI engine compared with the CI engine.

- At low loads the efficiency is about 10 percent, rising to about 25 percent at full load.

- The loss to coolant is about 60 percent at low loads and 30 percent at full load.

- The exhaust temperature rises very slowly with load and as mass flow rate of exhaust gas is reduced because the mass flow rate of fuel into the engine is reduced, the percentage loss to exhaust remains nearly constant (about 21% at low loads to 24% at full load).

- Percentage loss to radiation increases from about 7% at loads or 20% at full load.

3.9 Performance of CI Engines

The performance of a CI engine at constant speed variable load is shown in Fig. 3.13.

- As the efficiency of the CI engine is more than the SI engine the total losses are less. The coolant loss is more at low loads and radiation, etc., losses are more at high loads.

- The B_{mep}, BP and torque directly increase with load, as shown in Fig. 3.14. Unlike the SI engine BP and B_{mep} are continuously raising curves and are limited only by the load. The lowest brake specific fuel consumption and hence the maximum efficiency occurs at about 80 percent of the full load.

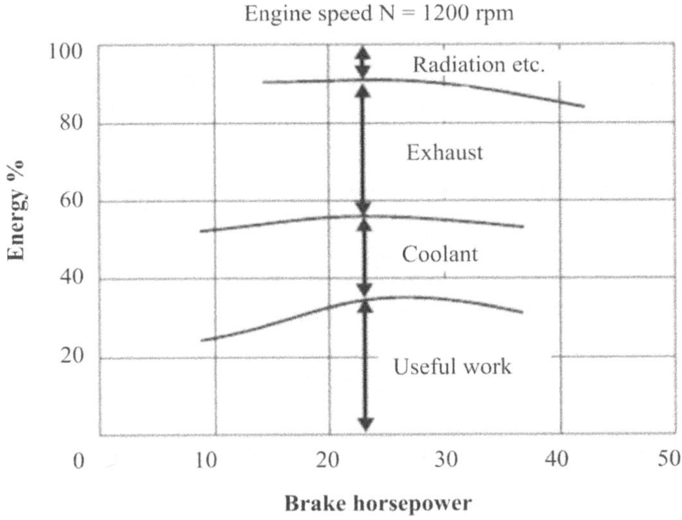

Fig. 3.13 Heat Balance Vs Load for a CI Engine

Fig. 3.15 shows the performance curves of variable speed GM 7850 cc four cycle V-6 Toro-flow diesel engine. The maximum torque value is at about 70 percent of maximum speed compared to about 50 percent in the SI engine. Though the bsfc is low, most of the speed range of the diesel engine is better than the SI engine.

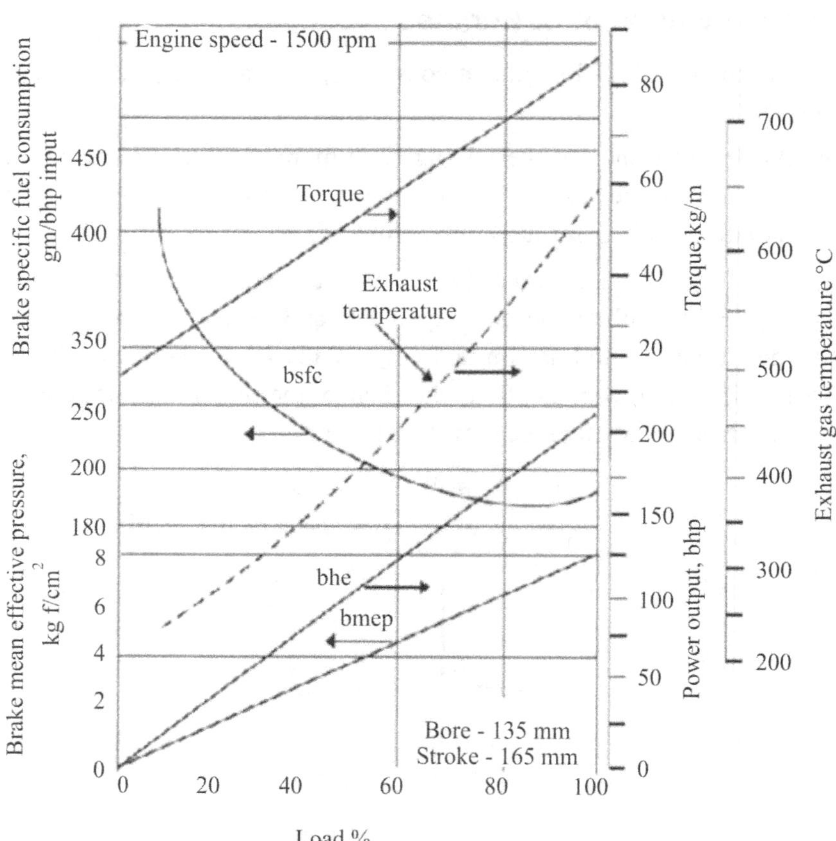

Fig. 3.14 Performance Curves of Six Cylinder Four – Stroke Cycle

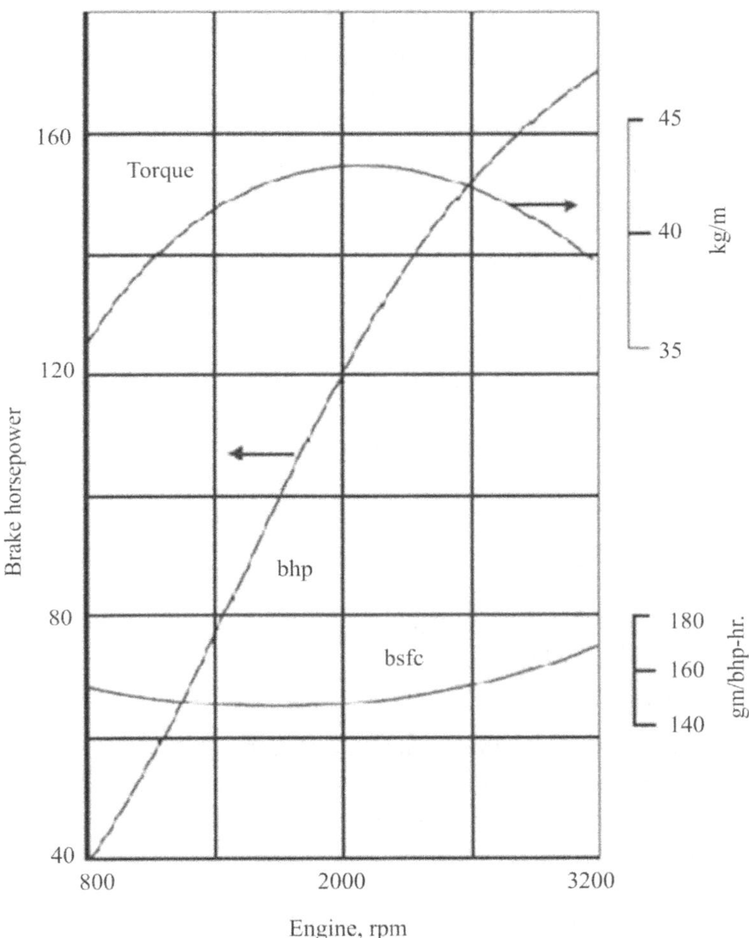

Fig. 3.15 Performance Curves of GM – Four Cycle Toro-flow Diesel Engine

Numerical Problems

Example 3.1: A gasoline engine works on Otto cycle. It consumes 8 litres of gasoline per hour and develops power at the rate of 25 kW. The specific gravity of gasoline is 0.8 and its calorific value is 44000 kJ/kg. Find the indicated thermal efficiency of the engine.

Solution: Heat liberated at the input

$$= m \, C_v$$

$$= 8 \times (0.8/60 \times 60)$$

$$= 6.4/3600$$

Power at the input

$$= (6.4/3600) \times 44000 \text{ kW}$$

$$\eta_{ith} = (\text{Output power/ Input power})$$

$$= \frac{25}{\dfrac{6.4 \times 44000}{3600}}$$

$$= \frac{25 \times 3600}{6.4 \times 44000} = 0.3196$$

$$= 31.96\%$$

Example 3.2: A single cylinder engine operating at 2000 rpm develops a torque of 8 N-m. The indicated power of the engine is 2.0 kW. Find loss due to friction as the percentage of brake power.

Solution:

$$\text{Brake powder} = \frac{2\pi NT}{60000} = \frac{2 \times \pi \times 2000 \times 8}{60000}$$

$$= 1.6746 \text{ kW}$$

$$\text{Friction power} = 2.0 - 1.6746$$

$$= 0.3253$$

$$\% \text{ loss} = \frac{0.3253}{2} \times 100$$

$$\% \text{ loss} = 16.2667\%$$

Example 3.3: A diesel engine consumes fuel at the rate of 5.5 gm/sec. and develops a power of 75 kW. If the mechanical efficiency is 85%, calculate bsfc and isfc. The lower heating value of the fuel is 44 MJ/kg.

Solution:

$$b\,sec = \frac{kW\ heat\ input}{kW\ heat\ output}$$

$$= \frac{C_v \times m_f}{P} = C_v \times bsfc$$

$$bsfc = \frac{5.55}{75} = 0.074\ g/kWs$$

$$= 0.074 \times 10^{-3} kg\,/\,kWs$$

$$C_v = 44\ MJ/kg = 44 \times 10^3 kJ\,/\,kg$$

$$b\,sec = bsfc \times C_v = 44 \times 10^3 \times 0.074 \times 10^{-3} = 3.256$$

$$i\,sec = b\,sec \times \eta_n = 3.256 \times 0.85$$

$$i\,sec = 2.7676$$

Example 3.4: Find the air-fuel ratio of a 4-stroke, 1 cylinder, air cooled engine with fuel consumption time for 10 cc as 20.0 sec and air consumption time for 0.1 m^3 as 16.3 sec. The load is 16 kg at speed of 3000 rpm. Also find brake specific fuel consumption in g/kWh and thermal brake efficiency. Assume the density of air as 1.175 kg/m^3 and specific gravity of fuel to be 0.7. The lower heating value of fuel is 44 MJ/kg and the dynamometer constant is 5000.

Solution:

$$Air\ consumption = \frac{0.1}{16.3} \times 1.175 = 7.21 \times 10^{-3} kg\,/\,s$$

$$Fuel\ consumption = \frac{10}{20} \times 0.7 \times \frac{1}{1000} = 0.35 \times 10^{-3} kg\,/\,s$$

$$Air\text{–}fuel\ ratio = \frac{7.21 \times 10^{-3}}{0.35 \times 10^{-3}} = 20.6$$

$$Power\ output\ (P) = \frac{WN}{Dynamometer\ constant}$$

$$= \frac{16 \times 3000}{5000} = 9.6\ kW$$

$$\text{bsfc} = \frac{\text{Fuel consumption (h/hr)}}{\text{Power output}}$$

$$= \frac{0.35 \times 10^{-3} \times 3600 \times 1000}{9.6}$$

$$\text{bsfc} = 131.25 \text{ g/k Wh}$$

$$\eta_{\text{bth}} = \frac{9.6}{0.35 \times 10^{-3} \times 44000} \times 100$$

$$\eta_{\text{bth}} = 62.3377$$

Example 3.5: A six-cylinder, gasoline engine operates on the four-stroke cycle. The bore of each cylinder is 80 mm and the stroke is 100 mm. The clearance volume per cylinder is 70 cc. At the speed of 4100 rpm, the fuel consumption is 5.5 gm/sec. [or 19.8 kg/hr.] and the torque developed is 160 Nm.

Calculate: (i) Brake power, (ii) The brake mean effective pressure, (iii) Brake thermal efficiency if the calorific value of the fuel is 44000 kJ/kg and (iv) The relative efficiency on a brake power basis assuming the engine works on the constant volume cycle $\gamma = 1.4$ for air.

Solution:

$$\text{BP} = \frac{2\pi \text{NT}}{60000} = \frac{2\pi \times 4100 \times 160}{60000} = 68.66$$

$$\text{B}_{\text{mep}} = \frac{\text{BP} \times 60000}{\text{LAnK}}$$

$$= \frac{68.66 \times 60000}{0.1 \times \dfrac{\pi}{4} \times (0.08)^2 \times \dfrac{4100}{2} \times 6}$$

$$\text{B}_{\text{mep}} = 6.6 \text{ bar}$$

$$\eta_{\text{bth}} = \frac{\text{BP}}{\text{m}_f \times \text{C}_v} = 29.03\%$$

Compression ratio, $r = \dfrac{V_s + V_d}{V_d}$

$$V_s = \frac{\pi}{4}D^2 L = \frac{\pi}{4} \times 8^2 \times 10 = 502.65 \text{ cc}$$

$$r = \frac{502.65 + 70}{70}$$

$$r = 8.18$$

Air standard efficiency,

$$\eta_{otto} = 1 - \frac{1}{(8.18)^{0.4}} = 1 - \frac{1}{2.3179} = 0.56858$$

(iv)   Relative efficiency, $\eta_{rel} = \dfrac{0.2903}{0.568} \times 100 = 51.109\%$

$$\eta_{bth} = \frac{BP}{m_f \times C_v}$$

$$= \frac{119.82 \times 60}{\dfrac{4.4}{10} \times 44000} \times 100$$

$$\eta_{bth} = 37.134\%$$

Volume flow rate of air at intake condition.

$$a = \frac{6 \times 287 \times 300}{1 \times 10^5} = 5.17 \text{ m}^3 / \text{min}$$

Swept volume per minute,

$$V_s = \frac{\pi}{4}D^2 L \, n \, k$$

$$= \frac{\pi}{4} \times (0.1)^2 \times 0.9 \times \frac{4500}{2} \times 9$$

$$= 127.17 \text{ m}^3 / \text{min}$$

olumetric efficiency,

$$\eta_v = \frac{5.17}{127.17} \times 100$$

$$\eta_v = 4.654\%$$

Air-fuel ratio, $\dfrac{A}{F} = \dfrac{6.0}{0.44} = 13.64$

Example 3.6: A gasoline engine is specified to be 4-stroke and four-cylinder. It has a bore of 80 mm and a stroke of 100 mm. On test it develops a torque of 75 N-m when running at 3000 rpm. If the clearance volume in each cylinder is 60 cc the relative efficiency with respect to brake thermal efficiency is 0.5 and the calorific value of the fuel is 42 MJ/kg; determine the fuel consumption in kg/hr and the brake mean effective pressure.

Solution:

$$\text{Swept volume, } V_s = \frac{\pi}{4} \times 0.08^2 \times 0.1 = 5.024 \times 10^{-4} \, m^3/\text{cylinder}$$

$$= 502.4 \text{ cc/cylinder}$$

$$\text{Compression ratio} \quad = \frac{502.4 + 60}{60} = 9.373$$

$$\text{Air-standard efficiency} = 1 - \frac{1}{(9.373)^{0.4}} = 0.5914$$

$$\eta_{bth} = \text{Relative } \eta \times \text{Air-standard } \eta$$

$$= 0.5 \times 0.5914$$

$$= 0.2954$$

$$BP = \frac{2\pi NT}{60000} = \frac{2\pi \times 3000 \times 75}{60000} = 23.55 \, kW$$

$$\text{Heat supplied} \quad = \frac{23.55}{0.2957} = 79.64 \text{ kJ/s}$$

$$\text{Fuel consumption} = \frac{79.64 \times 3600}{42000} = 6.8264 \text{ kg/hr}$$

$$P_{bm} = \frac{P \times 60000}{V_s \, n \, K}$$

$$= \frac{23.55 \times 60000}{5.024 \times 10^{-4} \times \dfrac{3000}{2} \times 4} = 4.6875 \times 10^5 \, n/m^2$$

$$= 4.6875 \text{ bar}$$

Example 3.7: A six-cylinder, four-stroke engine gasoline engine having a bore of 90 mm and stroke of 100 mm has a compression ratio 8. The relative efficiency is 60%. When, the indicated specific fuel consumption is 3009 g/kWh, Estimate (i) The calorific value of the fuel and (ii) Corresponding fuel consumption given that I_{mep} is 8.5 bar and speed is 2500 rpm.

Solution:

$$\text{Air-standard efficiency} = 1 - \frac{1}{r^{r-1}} = 1 - \frac{1}{8^{0.4}} = 0.5647$$

$$\text{Relative efficiency} = \frac{\text{Thermal efficency}}{\text{Air - standard efficiency}}$$

Indicated thermal efficiency $= 0.6 \times 0.5647 = 0.3388$

$$\eta_{ith} = \frac{1}{i_{sfc} \times C_v}$$

$$C_v = \frac{1}{n_{ith} \times i_{sfc}} = \frac{3600}{0.3 \times 0.3388}$$

$$C_v = 35417.035 \text{ kJ/kg}$$

$$IP = \frac{P_{im} LANk}{60}$$

$$= \frac{8.5 \times 10^5 \times 0.1 \times \dfrac{\pi}{4} \times 0.09^2 \times \dfrac{2500}{2} \times 6}{60000} = 67.6 \text{ kW}$$

$$\text{Fuel consumption} = \text{isfc} \times IP = 0.3 \times 67.6$$

$$IP = 20.28 \text{ kg/h}$$

Example 3.8: The observations recorded after the conduct of a retardation test on a single-cylinder diesel engine are as follows:

Rated power $= 10$ kW

Rated speed $= 500$ rpm

Sl. No.	Drop in speed	Time for fall of speed at no load, t_2(s)	Time for fall of speed at 50% load, t_3(s)
1.	500-400	7	2.2
2.	500-350	10.6	3.7
3.	500-325	12.5	4.8
4.	500-300	15.0	5.4
5.	500-275	16.6	6.5
6.	500-250	18.9	7.2

Determine the torque at full load and half load, friction power and mechanical efficiency.

Solution: First we draw a graph of drop in speed versus time taken for the drop.

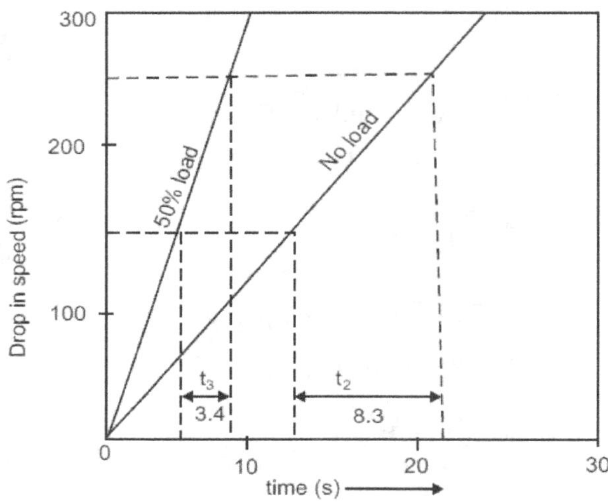

Fig. 3.16 Speed vs Time

$$P = \frac{2\pi NT}{60000}$$

Full load torque, $\quad T = \dfrac{P \times 60000}{2\pi N} = \dfrac{10 \times 60000}{2 \times \pi \times 500} = 191.083 \text{ Nm}$

Torque at half load, $\quad T_{1/2} = 95.5415 \text{ Nm}$

From graph:

Time for the fall of 100 rpm at no load, $t_2 = 8.3$ sec.

Time for the fall of same 100 rpm at half load, $t_3 = 3.4$ sec.

$$T_f = \frac{t_3}{t_2 - t_3} \times \text{Torque at 50\% load}$$

$$= \frac{t_3}{t_2 - t_3} \times T_{1/2}$$

$$= \frac{3.4}{(8.3 - 3.4)} \times 95.5415 = 66.294 \text{ Nm}$$

$$\text{Friction power} = \frac{2\pi NT_f}{60000} = \frac{2\pi \times 500 \times 66.294}{60000} = 3.469 \text{ kW}$$

$$\text{Mechanical efficiency } \eta_m = \frac{BP}{BP + FP} = 74.24\%$$

Example 3.9: A 4-cylinder, 4-stroke cycle engine having cylinder diameter 100 mm and stroke 120 mm was tested at 1600 rpm and the following readings were obtained.

Fuel consumption = 0.27 litres/minute, Specific gravity fuel = 0.74, B.P. = 31.4 kW, Mechanical efficiency = 80%, Calorific value of fuel = 44000 kJ/kg.

Determine:

   (i)   bsfc   (ii) I_{mep} and   (iii) Brake thermal efficiency.

Solution:

$$D = 100 \text{ mm} = 0.1 \text{ m}$$

$$L = 120 \text{ mm} = 0.12 \text{ m}$$

$$\eta_m = 80\% = 0.8$$

  (i) Brake specific fuel consumption (bsfc):

$$= \frac{0.27 \times 0.74 \times 60}{31.4} = 0.38174 \text{ kJ/kW.hr}$$

 (ii) Indicated power:

$$I.P = \frac{n \times P_{imp} \times L \times A \times N}{2 \times 60}$$

$$\frac{B.P}{\eta_m} = \frac{n \times P_{imp} \times 0.12 \times \frac{\pi}{4} \times (0.1)^2 \times 1600}{2 \times 60}$$

$$\therefore \qquad \frac{31.4}{0.8} = 4 \times P_{imp} \times 0.01256637$$

$$\therefore \qquad P_{imep} = 780.85 \text{ kN/m}^2$$

 (iiii)   Brake thermal efficnecy:

$$\eta_{bth} = \frac{\text{Brake power}}{\text{Heat supplied}}$$

$$= \frac{31.4}{\dfrac{0.27 \times 0.74}{60} \times 44000} \times 100 = 21.43\%$$

Example 3.10: A single cylinder and stroke cycle I.C. engine when tested, the following observations available:

Area of indicator diagram = 3 sq.cm, Length of indicator diagram = 4 cm, Spring constant = 10 bar/cm, Speed of engine = 400 rpm, Brake drum diameter = 120 cm, Dead weight on brake = 380 N, Spring balance reading = 50 N, Fuel consumption = 2.8 kg/hr, C_v = 42000 kJ/kg, Cylinder diameter = 16 cm, Piston stroke = 20 cm.

Find:

   (i)   F.P.

  (ii)   Mechanical efficiency

 (iii) bsfc, and

 (iv) Brake thermal efficiency.

Solution: Indicated mean effective pressure,

$$P_{imp} = \frac{\text{Area of indicate diameter}}{\text{Length of indicated diameter}} \times \text{Spring constant}$$

$$= \frac{A_i}{L_i} \times K_i$$

$$= \frac{3}{4} \times 10$$

$$P_{imep} = 7.5 \text{ bar}$$

Indicated power

$$I.P = \frac{P_{imp} \times L \times A \times N}{60}$$

$$= \frac{7.5 \times 10^5 \times 0.2 \times \frac{\pi}{4}(0.16)^2 \times 400}{60 \times 2} = 10.05 \text{ kW}$$

Brake power

$$B.P = \frac{2\pi NT}{60} = \frac{2\pi N(W - S)\frac{b}{2}}{60}$$

$$B.P = \frac{2\pi \times 400(380 - 50)}{60} \times \frac{1.2}{2} = 8.294 \text{ kW}$$

   (i)   Frictional power = F.P. = I.P – B.P.

$$= 10.05 - 8.294$$

$$= 1.756 \text{ kW}$$

(ii)　Mechanical efficiency $\eta_m = \dfrac{B.P}{I.P} = \dfrac{8.294}{10.05} \times 100 = 82.53\%$

(iii)　Brake thermal fuel consumption (bsfc):

$$= \frac{2.8}{8.294} = 0.3376 \text{ kg/kWh}$$

(iv)　Brake thermal efficiency (η_{bth})

$$\frac{B.P}{Heat\ supplied} = \frac{8.294}{\dfrac{2.8}{3600} \times 42000} \times 100 = 25.39\%$$

Example 3.11: A six-cylinder 4-stroke petrol engine having a bore of 90 mm and stroke of 100 mm has a compression ratio of 7. The relative efficiency with reference to indicated thermal efficiency is 55% when indicated mean specific fuel consumption is 0.3 kg/kWh. Estimate the calorific value of the fuel and fuel consumption in kg/hr. Given that indicated mean effective pressure is 8.5 bar and speed is 2500 rpm.

Solution:

Number of cylinders (n) = 6, L = 100 mm = 0.1 m

D = 90 mm = 0.09 m, r = 7

$\eta_t = 55\% = 0.55$ [based on indicated thermal efficiency]

isfc = 0.3 kg/kWh

$$P_{mi} = 8.5 \text{ bar}$$

$$N = 2500 \text{ rpm}$$

$$I.P = \frac{P_{imp} \times A \times L \times N}{60000} \times n_1 \qquad\qquad .....(1)$$

Where, $n = \dfrac{N}{2} = \dfrac{2500}{2} = 1250$ strokes/m [for 4 stroke engine]

$$A = \frac{\pi}{4} d^2$$

From eq. (1), we have,

$$I.P = (8.5 \times 10^5) \times \frac{\pi}{4} (0.09)^2 \times 0.1 \times 1250 \times 6 \times \frac{1}{60000}$$

$$= 67.593 \text{ kW}$$

(i)   Fuel consumption, m_f :

$$\text{isfc} = \frac{\text{Fuel Consumption}\left(\dfrac{kg}{hr}\right)}{\text{Indicated Power}}$$

$0.3 = m_f/67.593$

$m_f = 20.278 \text{ kg/hr}$

(ii)   Calorific Value (C_v) of fuel:

Air standard efficiency

$$\eta_a = 1 - \frac{1}{(r)^{(\gamma-1)}} = 1 - \frac{1}{(7)^{(1.4-1)}} = 0.42467$$

Relative efficiency, $\eta_r = \dfrac{\text{Indicate thermal efficiency, } \eta_i}{\text{Air standard efficiency, } \eta_a}$

$$\eta_i = \eta_r \times \eta_a = 0.55 \times 0.42647 = 0.2346$$

But

$$\eta_i = 0.2346 = \frac{67.593}{\left(\dfrac{20.278}{3600}\right) \times C_v}$$

$\therefore$ $C_v = 51150.6 \text{ kJ/kg}$

Example 3.12: The following observations were recorded during a trial on a 4-stroke diesel engine:

Power absorbed by non-firing engine when driven by an electric motor = 10 kW

Speed of the engine = 1750 rpm

Brake torque = 327.4 Nm

Fuel used = 15 kg/hr.

Calorific value of fuel = 42000 kJ/kg

Air supplied = 4.75 kg/min

Cooling water circulated = 16 kg/min

Outlet temperature of cooling water = 65.8 °C

Temperature of exhaust gas = 400 °C

Room temperature = 20.8 °C

Specific heat of water = 4.19 kJ/kg·K

Specific heat of exhaust gas = 1.25 kJ/kg·K

Determine:

    (i)   BP

   (ii)   Mechanical efficiency

 (iii)   bsfc

 (iv)   Draw up heat balance sheet on kW basis

Solution:

   (ii)   Brake power (BP):

$$BP = 2\pi\, NT = 2 \times \pi \times \frac{1750}{60} \times 327.4 \times 10^{-3}$$

$$= 60.01 \text{ kW}$$

$$\text{Mechanical efficiency } (\eta_m) = \frac{BP}{IP}$$

But   FP = IP – BP

Given that power absorbed by non-firing engine when driven by electric motor. This is frictional power.

$$F.P. = 10 kW$$

This type of testing is done in a monitoring test which is used to calculate the frictional power of an engine.

Hence,   FP = 10 kW

$\therefore$ IP = BP + FP

$$= 60.01 + 10$$

$\therefore$ IP = 70.01 kW

$$\eta_m = \frac{60.01}{70.01} = 0.8571 = 85.71$$

 (iii)   bsfc: Brake specific fuel consumption:

$$\therefore \quad bsfc = \frac{m_f}{BP} = \frac{15}{60.01} = 0.25 kg / lWh$$

 (iv)   Heat balance sheet in kW basis:

        (i)   Power supplied by fuel $= m_f \times C_v$

$$= \frac{15}{3600} \times 42000 = 175 kW$$

        (ii)   Brake power = 60.01 kW

       (iii) Power to cooling water $= m_w\, C_{pw}\Delta T$

$$= \frac{16}{60} \times 4.19 \times (T_o - T_{in})$$

$$T_o = 65.8 + 273 = 338.8 \text{ K}$$

$$T_{in} = 20.8 + 273 = 293.8 \text{ K}$$

Power lost to cooling water = 50.28 kW

(iv) Power to exhaust = $m_E \, C_{pE} \Delta T$

Here, mass of exhaust gases

$$m_E = m_a + m_f$$

$$= \frac{4.75}{60} + \frac{15}{3600} = 0.0833 \text{ kg/s}$$

Power lost to exhaust = $0.0833 \times 1.25 \times (400 - 20.8)$

$$= 39.48 \text{ kW}$$

Heat balance sheet:

	Input (kW)	%	Output	kW	%
01	Power from fuel 175 kW	100%	Brake power	60.01	34.29
			Power lost cooling water	50.28	28.73
			Power lost to exhaust	39.48	22.56
			Unaccounted power	25.23	14.42
Total	174 kW	100%	Total	175	100%

CHAPTER 4

Combustion in SI Engines

- Spark ignition engine mixture requirements
- Feed back control carburetors
- Stages of combustions
    - (i) Normal combustion
    - (ii) Abnormal combustion
- Fuel Injection systems
    - (i) Mono point injection
    - (ii) Multi point injection
- Factors affecting knock
- Combustion Chambers

4.1 Mixture Requirements of Automotive Engines

The simple carburetor must be fitted with additional systems to provide different air-fuel ratio for automotive engines operating under variable speed and load. There are basically three steady ranges of operation for an automotive engine.

1. Idling+ (mix must be enriched)
2. Cruising (mix must be leaned)
3. High power (mix must be enriched)

***For average cruising speeds*:** The air fuel ratio is approximately from 15:1 to 17:1.

***To obtain maximum power*:** To accelerate the engine quickly a richer mixture of the ratio of about 12:1 is desirable. This is also called maximum power ratio. To start the engine from cold even a mixture richer than this but above the lower limit is obtained.

***For maximum economy*:** i.e., less fuel consumption per unit power the mixture ratio may be approximately 16:1 to 17:1 as shown in Fig.4.1.

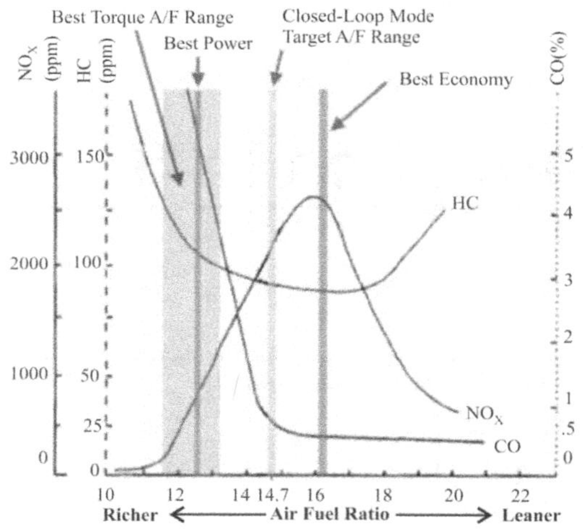

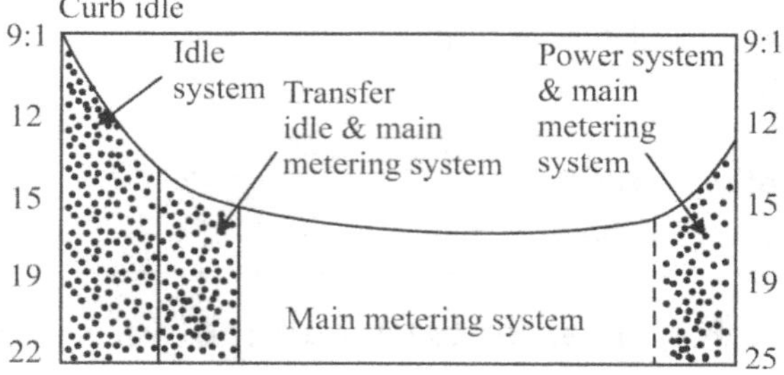

Fig. 4.1 Load (A/F Mixture Requirements)

Mixture Requirements for Steady State Operation

(i) ***Idling and low speed (from no load to about 2 percent of rated power)*:** Idling refers to no power demand. During idling air supply is throttled and the residual gases make up a large fraction of the charger at the end of the suction period. In addition, during valve overlap period some exhaust gases are drawn back into the cylinder. The result is a chemically correct (stoichiometric) mixture of air and fuel (=15:1) would be so diluted by residual gases that combustion would be erratic or impossible. A richness should gradually change to slightly lean for the second range.

(ii) ***Normal power (from about 25 percent to about 75 percent of rated power)***: In the normal power range the main consideration is fuel economy. Because mixture of fuel and air is never completely homogeneous the stoichiometric mixture of fuel and air will not burn completely and some fuel will be wasted. For this reason excess of air, say 10% above theoretically correct (=A/F 16.5:1), is supplied in order to ensure complete burning of the fuel.

(iii) ***Maximum power (from 75% to 100% of rated power)***: Maximum power obtained when all the air supplied is fully utilized. As the mixture is not completely homogeneous a rich mixture must be supplied to assure utilization of all air (though this would mean wasting some fuel, which would pass in exhausts in unburned state). The air fuel ratio for maximum power is about 13:1.

A.   Transient mixture requirements: In addition to the steady mixture requirements discussed above, there are mixture requirements of transitory nature during starting up and acceleration.

(i) *Starting up*: To start the engine from cold, an even richer mixture than that for idling is required because part of the fuel supplied remains in the liquid state, and some of the vaporized fuel may condense on cold cylinder walls and piston heads. Although the air ratio at the carburetor may be rich enough for satisfactory combustion of the mixture the ratio of evaporated fuel to air in the cylinder may be far too lean to ignite and burn. It may be necessary to supply temporarily as much as 5 or 10 times the normal amount of fuel at the carburetor to obtain the correct mixture strength in the cylinder until the manifold and cylinder parts become warm.

(ii) *Acceleration*: When an engine is to be accelerated it is necessary to supply an extra quantity of fuel to enrich the mixture temporarily to produce rapid and smooth acceleration.

4.2   Computer Controlled Carburetor

Computer-controlled carburetors also called as Feedback controlled carburetors are one in which air-fuel ratio mixture is automatically controlled by a computer operated solenoid dithering valve. The feedback carburetor system provides positive air/fuel ratio control for maximum reduction of emissions. The Electronic Control Unit (ECU) receives signals from various sensors and then modulates 2 solenoid valves installed on the carburetor to control the air/fuel ratio. The ECU also controls the ignition timing, electric choke and the idle up solenoid by switching the solenoid valves ON and OFF.

Working: The air/fuel ratio control is maintained by the ECU in 1 of 2 operating modes as shown in Fig.4.2.

Closed Loop Control (Feedback Control)

Closed loop control is used after engine warm-up. The air/fuel mixture is determined by the feedback control, based on the oxygen sensor signal. Oxygen sensor output voltage changes sharply at the stoichiometric ratio (14.7:1). The control unit senses this signal and uses the feedback solenoid valve to regulate the ratio. By providing a stoichiometric ratio, the best purification rate of the 3-catalyst converter may be kept. In this state, the slow cut solenoid valve is kept wide open (100 percent duty).

Open Loop Control

During engine start, warm-up, high load operation and deceleration, the air/fuel ratio is in open loop. Control is based on preset values contained in the ECU. During deceleration, the slow cut solenoid valve limits fuel flow for better fuel economy and for prevention of overheating of the catalysts.

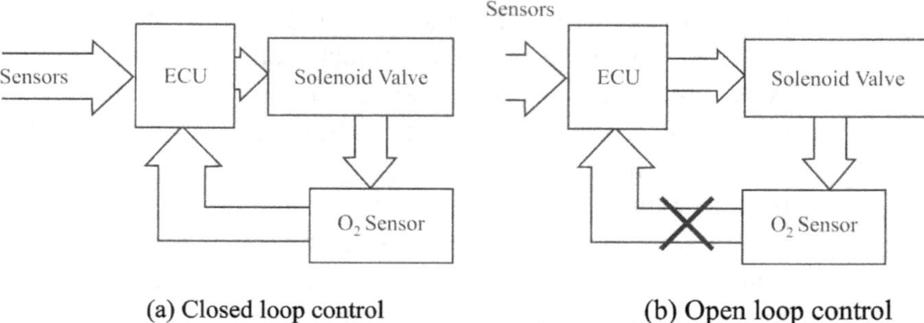

(a) Closed loop control   (b) Open loop control

Fig. 4.2   Feedback Carburetor System

Components of Computer-Controlled Carburetors

Electronic Control Unit (ECU)

The electronic control unit is mounted in the passenger compartment and consists of a printed circuit board mounted in a protective metal box. It receives analog inputs from sensors and converts them into digital signals. These signals and various inputs are processed and used by the ECU in controlling the fuel delivery, secondary air, deceleration spark and throttle opener management.

Coolant Temperature Sensor

The coolant temperature sensor is installed in the intake manifold. This sensor provides data to the ECU for use in controlling fuel delivery and secondary air management. The temperature sensor detects the operating temperatures of the engine. Its resistances changes with the temperatures. This allows the computer to enrich the air fuel mixture during cold engine operation.

Pressure Sensor

Pressure sensor measures intake manifold vacuum and engine load. High engine load and power output causes intake manifold vacuum to drop. The pressure sensor, in this way can signal the computer with a change in resistance and current flow.

Throttle Position Sensor

The Throttle Position Sensor (TPS) is a potentiometer mounted to the carburetor. The TPS provides throttle angle information to the ECU to be used in controlling the fuel delivery and secondary management.

Engine Speed Sensor

The engine speed sensor signal comes from the negative terminal voltage of the ignition coil. Electric signals are sent to the ECU, where the time between these pulses is used to calculate engine speed, which is used in controlling fuel delivery, secondary air management, deceleration spark and throttle opener management. An idle speed actuator may also be used to allow the computer to change engine idle speed. It is normally a tiny electric motor and gear mechanism that holds the carburetor throttle lever in the desired position.

Oxygen Sensor

The oxygen sensor is mounted in the exhaust manifold. The output signal from the sensor, which varies with the oxygen content of the exhaust gas stream, is provided to the ECU for use in controlling closed loop compensation of fuel delivery. An oxygen sensor or exhaust gas sensor monitors the oxygen content in the engine richness (low oxygen content) or any change in oxygen content in the exhaust.

Vacuum Switch

The switch is installed on the floor board and is electronically turned ON (switch closed) when the throttle valve is in the closed (idling) position. Information to this switch is provided from the ECU for use in controlling fuel delivery and secondary air management.

Feedback Solenoid Valve

The feedback solenoid valve is installed in the carburetor float chamber cover. The ECU controls the air to fuel ratio by controlling the duty cycle of the valve. As the duty ratio increases, the mixture becomes leaner. Computer controlled carburetor normally uses a Plunger Shaped solenoid valve to open and close the air fuel passages in the carburetor.

1. When the solenoid is turned on, the valve is lifted which increases the fuel supply through the jet.

2. When the solenoid is turned off, the spring pushes the valve down to decrease the fuel supply.

3. The solenoid is operated by a computer, according to the signals received by it on engine speed, coolant temperature, manifold vacuum, throttle position and the amount of oxygen going out in the exhaust.

4. Due to continual up and down movement of the plunger valve, it is called a dithering valve.

4.3   Stages of Combustion of SI Engines

Combustion stages in S.I. engine can be divided into three categories as shown in Fig.4.3.

1. Normal combustion
2. Pre ignition
3. Detonation

Normal Combustion

A high temperature spark is produced by a spark plug. This spark travels through the air fuel mixture. This sparks leaves a thin thread of flame behind it. The air fuel mixture enveloped around the thin thread of flame gets ignited and combustion commences. Since the air fuel mixture is in turbulent condition, the surface area of heat transfer is more and combustion is speeded up enormously. In P-□ diagram we can see the stages of normal combustion. ABC is normal compression curve.

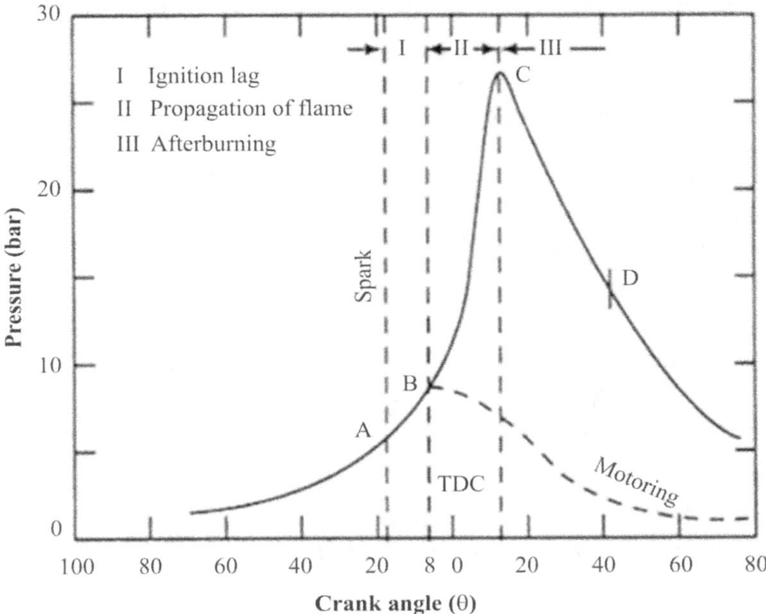

Fig. 4.3 Stages of Combustion in SI Engine

The first stage **(A-B)** is referred to as the **ignition lag or preparation phase** in which growth and development of a self propagating nucleus of flame takes place. This is chemical process depending upon both temperature and pressure, the nature of the fuel and the proportion of the exhaust residual gas. Further, it also depends upon the relationship between the temperature and the rate of reaction.

The second stage **(B-C)** is a physical one and it is concerned with the spread or **Propagation of the flame** throughout the combustion chamber. The starting point of the second stage is where the first measurable rise of pressure is seen on the indicator diagram i.e., the point where the line of combustion departs from the compression line (Point B). This can be seen from the deviation from the motoring curve.

During the second stage the flame propagates practically at a constant velocity. Heat transfer to the cylinder wall is low because only a small part of the burning mixture comes in contact with the cylinder wall during this period. The rate of heat-release depends largely on the turbulence intensity and also on the reaction rate which is dependent on the mixture composition. The rate of pressure rise is proportional to the rate of heat-release during this stage, the combustion chamber volume remains practically constant (since piston is near the top dead center).

The starting point of the **third stage** is usually taken as the instant at which the **maximum pressure** is reached on the indicator diagram (point C). The flame velocity decreases during this stage. The rate of combustion becomes low due to lower flame velocity and reduced flame front surface. Since the expansion stroke starts before this stage of combustion, with the piston moving away from the top dead center, there can be no pressure rise during this stage.

Ignition Lag: The time period between first igniting fuel and commencement of main phase of combustion is called ignition lag (or) period of incubation. The ignition lag is normally 0.0015 sec.

4.4   Effect of Various Engine Variables on Ignition Lag

The process during ignition lag is a chemical one and depends on the following factors.

1. *Compressed gas temperature and pressure*: The rate of chemical reaction is very small at low temperature but increases rapidly at high temperatures. The rate of chemical reaction also depends on density or pressure, but to a small extent. The ignition lag therefore decreases by increasing the temperature and pressure of the gas at the time of spark. Thus, increasing the compression ratio and intake temperature and pressure, and retarding the spark timing tend to reduce the ignition lag.

2. *Mixture ratio*: The ignition lag is smallest for the mixture ratio which gives the maximum temperature. This mixture ratio is somewhat richer (about 10%) than the stiochiometric ratio.

3. *Fuel*: The ignition lag depends on the chemical nature of the fuel. The higher the self-ignition temperature of the fuel, the longer is the ignition lag.

4. *Speed or turbulence*: Turbulence is directly proportional to engine speed. Increasing engine speed does not affect much as the ignition lag is measured in milliseconds. However, it is measured in degrees of crank rotation, the ignition lag increases almost linearly with engine speed. For this reason it becomes necessary to advance the spark timing at higher speed. Excessive turbulence in the area of spark plug is harmful because it increases the heat transfer from the combustion zone and leads to unstable development of the nucleus of flame. That is why the spark plug is usually arranged in a small recess in the wall of the combustion chamber.

5. ***Electrode gap***: The electrode gap is important from the point of view of establishment of the nucleus of flame. If the gap is too small, quenching of the flame nucleus may occur and the range of fuel-air ratio for the development of the flame nucleus is reduced. Lower compression ratio and lighter load requires more electrode gap. For compression ratio of a gap of 0.625 mm is satisfactory.

4.5   Effects of Engine Variables on Flame Propagation

A study of the variables which affect the flame propagation velocity is important because the flame velocity influence the state of pressure rise in the cylinder and has bearing on certain types of abnormal combustion. The various factors which affect the flame speed are discussed below, the most important factors being fuel-air ratio and turbulence.

1. ***Fuel-air ratio***: The maximum flame speed occurs when the mixture is about 10% richer than stoichiometric (A/F=13.5). Lean mixtures release less thermal energy resulting in lower flame temperature and flame speed. Very rich mixtures have incomplete combustion (some carbon only burns to CO and not to CO_2), which also results in production of least thermal energy and hence flame speed is again low.

2. ***Compression ratio***: An increase in compression ratio increases the temperature and density of gas at the end of compression, which reduces mixture preparation time. This makes combustion faster and thus flame speed increases. The compression ratio should be as high as possible to limit that there is no detonation.

3. ***Intake temperature and pressure***: Increase in intake temperature and pressure increases the flame speed as for the same compression ratio, temperature and pressure before compression increases.

4. ***Engine load***: With increase in engine load cycle pressures increase. Hence the flame speed increases.

5. ***Turbulence***: The flame speed is very low in non-turbulent mixtures. A turbulent motion of the mixture intensifies the process of heat transfer and mixing of the burned and unburned portions in the flame front (diffusion). These two factors cause the velocity. The turbulent of the mixture is due to admission of fuel-air mixture through comparatively narrow sections of the intake pipe, valves, etc. In the suction stroke, the turbulence can be increased at the end of the compression stroke by suitable design of combustion chamber which involves the geometry of

cylinder head and piston crown. The degree of turbulence increases directly with the piston speed.

6. ***Engine speed***: The higher the engine speed, the greater is the turbulence inside the cylinder. For this reason the flame speed increases almost linearly with engine speed. Thus if the engine speed is doubled the time in milliseconds required for the flame to traverse the combustion space would be halved. This, however, means that the number of crank degrees required for flame propagation will remain almost constant at all speeds.

7. ***Engine size***: Engines of similar design generally run at the same piston speed. This is because smaller engines run at higher r.p.m. And larger engines run at lower r.p.m. Due to the same piston speed, inlet velocity and degree of turbulence, flame speeds are nearly same in similar engines regardless of the size. However, in large engines the flame travel is longer. Therefore, if the engine size is doubled the time required in milliseconds for propagation of flame through combustion space will also be doubled. Since the bigger engine will be running at half the r.p.m. of the smaller engine, the number of crank degrees required for flame travel will be about the same irrespective of engine size.

Pre-ignition: A very high temperature carbon deposits formed inside the combustion chamber ignites the air fuel mixture before normal ignition occurred by spark plug. Sudden and violent noise experienced inside the engine cylinder is known as detonation, Pinking or Knocking or Hammer blow.

4.6   SI Engine Combustion Chamber and its Various Types

Requirements of a good combustion chamber for SI engines. The basic requirements of a good combustion chamber are high power output high thermal efficiency and smooth engine operation.

A. *High power output:* It requires

1. High volumetric efficiency

2. Maximum utilization of air

3. High compression ratio

4. Optimum degree of turbulence

B. *High thermal efficiencies:* It requires

1. High compression ratio
2. Small heat loss during combustion
3. Good scavenging of the exhaust gasses

C. *Smooth engine operation:* It requires

1. Moderate rate of pressure rise
2. Absence of detonation

Combustion Chamber Design

The following would be desirable characteristics of combustion chambers for spark ignition engines to avoid the tendency to detonate when operating under specified conditions.

1. Small bore.
2. Very high velocity through inlet valves.
3. Short Surface/volume ratio of flame path to bore (Spark plug should be located at the geometric symmetry of the combustion chamber).
4. Absence of hot surfaces away from the spark plug i.e., the regions where the gas is likely to detonate.
5. Use of squish areas, particularly in the region of the end gas i.e., the gas away from spark plug which is likely to detonate.

Types of Combustion Chambers

1. I head combustion chamber (or) Overhead Valve
2. L-head (or) Side-valve type Bath tub form
3. F-head combustion chamber
4. T-head combustion chamber
5. Hemispherical type
6. Stratified charge type
7. Multi-valve type

1. Overhead Valve (or) I head combustion chamber: In overhead valve design both the valves are in the head. The important advantages are shorter flame travel, higher volumetric efficiency due to larger valves and more direct passage ways, lesser heat loss due to lower surface-volume ratio and less force on the head due to smaller area. But it has complicated design valve mechanism.

*Bath tub form***:** This type of combustion chamber consists of oval shaped chamber with both valves mounted overhead. The spark plug is mounted at the side.

***Wedge form*:** This type of combustion chambers also consists of oval shaped chamber with both valves mounted overhead at its side with slight inclination. The spark plug is mounted centrally. The valves are inclined. It gives good performance.

2. **F-Head Combustion Chamber:** In this type of combustion chambers one valve is in overhead and other is in engine block. In this type plug cooling is efficient. The effective flame travel distance is less. This design gives high volumetric efficiencies maximum, compression ratio for fuel and high thermal efficiency. No misfiring is occurred.

3. **T-Head Combustion Chamber:** This was the earliest type of combustion chamber used by Ford Motor Corporation in the beginning of this century.

This head has the following disadvantages.

   (i) The inlet valve and the exhaust valve are separately actuated by two cam shafts driven from the crank shaft by gears.

   (ii) The flame path from the spark plug has to be large. And in order to compensate for the hot region near the exhaust valve the spark plug has to be located near the exhaust valve. This further increased the flame path in the combustion chamber from the spark plug to the end gas near the wall towards inlet valve. It was reported that the engine would detonate at compression ratio of 4 with a fuel of octane number of 50.

4. **L-Head or Side valve Combustion Chamber:** This shape was quite popular till the World War I. Its main advantages are:

   (i) Valves are placed in the block and therefore easy to enclose and lubricate the valve mechanism.

   (ii) The head is removable without disturbing the valve gear and main pipe work and therefore maintenance is easy.

In this combustion chamber, the spark plug is located near the exhaust valve, mostly over the exhaust valve in the form as shown in Fig.4.4. The performance was not very satisfactory because of the following shortcomings: (i) Long and arduous path for inlet air to reach the cylinder by passing through two bends leading to loss of velocity head and fall in turbulence level.

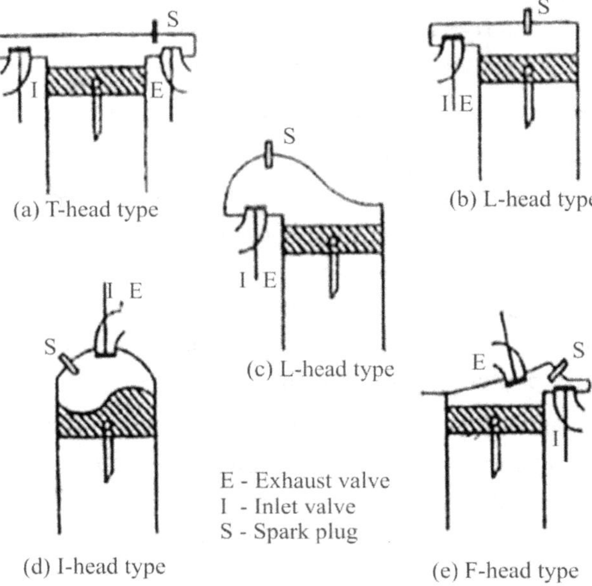

Fig. 4.4 Types of Combustion Chambers

Due to these limitations the side valve engines did not use higher compression ratios as compared to overhead valves which also started being used during the later period. Thus the side valve engines could not compete with over head valve engine in power and efficiency.

4.7 Detonation or Knock in SI Engines

In normal combustion the flame started by the spark travels across the combustion chamber in a fairly even way. Under certain engine operating conditions abnormal combustion may take place which is detrimental to the life and performance of the engine. There are a variety of abnormal combustions with complex phenomenon. The important abnormal combustion is detonation or knock and pre-ignition. Detonation or knock refers to the condition in which an audible knock is heard in the engine. Detonation is due to the propagation of a high speed pressure wave created by the sudden explosion of the end charge even before the flame ignites. Pre-ignition refers to ignition by a hot spot before the passage of the spark. The phenomenon of detonation is most important because it puts a limit on the compression ratio at which an engine can be operated, which in turn controls the efficiency and to some extent power output.

Though the phenomenon of detonation or knock has been under investigation for nearly last fifty years the exact mechanism of knocking is still not fully understood. There are two general theories of detonation:

(a) the auto-ignition theory, and (b) the detonation theory.

A.   Auto ignition theory: It is evident that if a petrol air mixture is compressed sufficiently it will ignite spontaneously. In a spark ignition engine the mixture is not normally raised to such a high temperature on the compression stroke that auto ignition or self-ignition of the charge will take place. Hence electric spark is necessary to initiate combustion. However, combustion mixtures will ignite spontaneously if conditions are suitable.

To understand the auto ignition theory of knock in SI engines let us compare normal combustion and knocking combustion shown Fig 4.5. Normal flame front travels across the combustion chamber from A towards D. The speed of the flame front is about 15-30 m/s. As the flame front advances it compresses the unburned charge BBD raising its temperature.

The temperature is also increased by radiation from the advancing flame from if this unburned charge does not reach its self-ignition temperature, it will not auto-ignite and the flame from BB will move across through the unburned charge to the farthest point of chamber D in the normal manner. The pressure crank angle diagram for normal combustion is a smooth curve and the rate of pressure rise (dp/dt) is within limits.

In the abnormal combustion called detonation or knocking, the end charge auto ignites before the flame from reaches it. In order to auto-ignite the last unburned portion of the charge must reach self-ignition temperature and must remain at this temperature for a certain length of time. Auto-ignition does not occur spontaneously at self ignition temperature but requires a finite period called delay period in which certain chemical reactions take place which prepare the charge for auto-ignition. In this case, there is a possibility of detonation or knocking. If the flame front can proceed from BB to D and consume the unburned charge in a normal manner prior to completion of delay period there will be no detonation. If however the flame front is able to proceed only as for as say CC during the ignition delay period then the remaining portion of the unburned change CCD will auto ignite setting high flame velocity

(2000-3000 m/s) and cause extreme pressure fluctuations. This gives rise to a pressure wave which strikes the cylinder wall and sets generating a vibrating characteristic pinking or ringing sound, as if struck by a light hammer.

Detonation of about 5 percent of total charge is sufficient to produce a violent knock. It should be noted that the maximum pressure of cycle may or may not greatly increase in detonation. It is the extremely high rate of pressure rise (dp/dt) that can physically damage the engine.

It is clear from the above description of the phenomenon of detonation refers to phenomenon associated with the sudden acceleration of a flame front. The detonation theory assumes that as the flame front progress at a moderate rate through the mixture, a critical condition is sometimes reached at which a violent pressure wave. i.e., a shock wave is formed. The critical condition is perhaps caused by the concentration of free radicals such as hydrogen atoms during the pre flame reactions. The violence of the shock wave and they traverse the cylinder together at supersonic velocity. The rapidly moving form is termed a detonation wave. The wave is reflected and re-reflected by the cylinder walls giving rise to knocking noise.

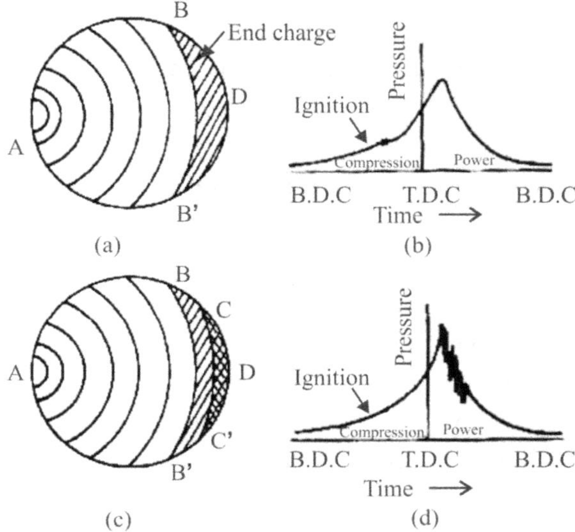

Fig. 4.5 Combustion with Normal and Detonation

Illustration of normal combustion, knocking and pre ignition are shown in Fig.4.6.

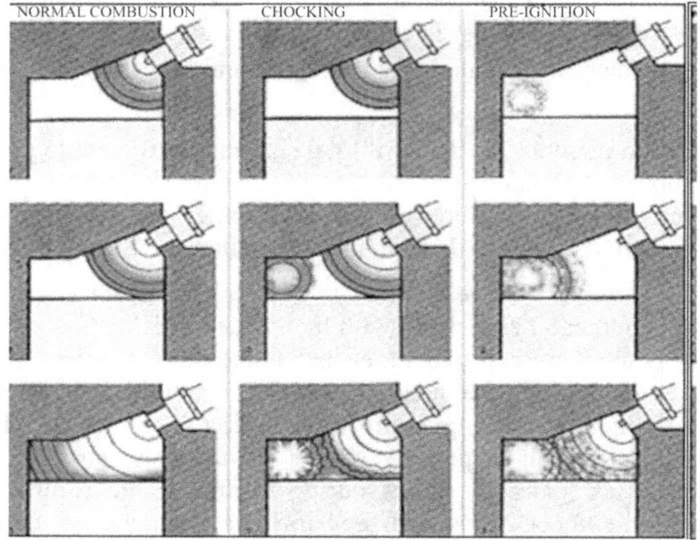

Fig. 4.6 (a) Normal Condition (b) Detonation (c) Pre-ignition

Effects of detonation

1. Undesirable noise and roughness
2. Mechanical damage to piston and connected components
3. Carbon deposits
4. Increase in heat transfer
5. Decrease in power output and efficiency
6. Pre-ignition
7. Heavy vibration of engines
8. Stresses produced in the material is higher
9. Inefficient combustion
10. Local overheating

Control of detonation

1. Increasing engine speed
2. Retarding spark (time factor)
3. Reducing pressure in the inlet manifold by throttling
4. Making the ratio too lean or too rich preferably latter
5. Water injection increases the delay period as well as reduces the flame temperature
6. Use of high-octane fuel can eliminate detonation

4.8   Effect of Various Engine Variables on SI Engine Knock (Factors Affecting Knocking)

1. Fuel Characteristics

(a) *Molecular Structure*: The knocking tendency is markedly affected by the type of the fuel. The petroleum fuels usually consist of hydrocarbon of different molecular structure. The structure of the fuel molecule has enormous effect on knocking tendency. Increasing the carbon-chain increase knocking tendency and centralizing the carbon atoms decreases the knocking tendency. Unsaturated hydrocarbons have less knocking tendency than saturated hydrocarbons.

(b) *Self Ignition Temperature*: Fuels with higher self-ignition temperature are less detonating as end portion of the mixture will not auto-ignite easily.

(c) *Flame Velocity*: The flame velocity is the distance traveled by the flame in the combustion chamber in unit time. The flame velocity also depends upon the type of fuel used, A:F ratio and pressure, temperature, conditions of charge.

Higher flame speeds are desirable as the mixture will be consumed by the flame and there will be no auto-ignition.

2. Conditions of the Charge

(a) *A:F Ratio*: The flame velocity or burning rate of the fuel is highest near the stoichiometric towards the rich-side. Therefore as rich mixture has better anti knocking quality compared to lean mix.

(b) *Density of the Charge*: Throttling of an engine or reducing the suction pressure reduces the trend of knocking where the supercharging increases the knocking tendency.

(c) *Temperature of Charge*: Higher initial temperature causes easier auto-ignition of the end of the mixture and so is not desirable.

3. Compression ratio

The pressure and temperature at the end of compression increase with increase in compression. The tendency of knocking increases with compression ratio.

(a) *Dilution by Residual Gases*: Higher compression ratio decreases dilution by residual gases thus increases the rate of burning and hence tendency to knock.

(b) ***Valve Timing***: Late closing of inlet valve reduces the percentage of exhaust gases left in the cylinder and it also reduces the effective compression stroke.

All these factors have appreciable effect on knock tendency.

4. Ignition

(a) ***Location of the Spark Plug***: The spark should be located so as to reduce the length of flame travel to minimum. This decreases the combustion time to a minimum so that the flame may pass through the end of mixture before knocking occurs. It should be located away from the exhaust valve to prevent pre-ignition of the charge.

(b) ***Spark Timing***: The quantity of fuel burned per cycle before and after TDC position depends on the spark timing. The temperature of the charge increases by increasing the spark advance and it increases the rate of burning and does not allow sufficient time to the end mixture to dissipate the heat and increase knocking tendencies. The knocking can be reduced by retarding the spark but with consequent loss in thermal efficiency.

(c) ***Engine Speed***: The flame speed increases with the increase in engine speed due to increased turbulence. The compression of the charge occurs rapidly in high speed engine and reduces the delay period and thereby knock tendency is reduced. This is because, at high speeds, the flame front reaches the end of the cylinder burns even the last part of the charge more quickly before the temperatures of the end mixtures is below ignition temperature.

5. Combustion Chamber

(a) ***Shape of Combustion Chamber***: If the combustion chamber is more compact in its shape, it provides maximum cylinder heads because a cool combustion chamber wall is desirable for high compression without knock. Combustion chambers with highly polished surfaces tend to cause knocking more readily; this is due to poor heat transfer characteristics of the polished surface. Appreciable carbon deposits result in hotter surface and increase the knocking tendency.

6. Engine Cylinder

(a) ***Temperature of the Walls***: Cooler walls of the cylinder have fewer tendencies for knocking. Improper lining of the cylinder also increases the knocking tendency.

(b) *Size***:** The size of the cylinder (diameter) should be as small as possible because the flame travels in more with larger diameter cylinder. It is always preferable to have multi cylinder engine of smaller than single cylinder engine of large diameter.

(c) *Temperature of Cooling Water***:** The effect of jacket water temperature is similar to the effect of temperature. Raising the jacket water temperature increases the end gas temperature and increases tendency of knock.

7. Other Factors

*Engine load***:** The cylinder wall temperature increases with increasing load on engine and thereby increases the knocking tendency.

4.9 Summary of Variables Affecting Knocking in SI Engine

The various factors affecting knocking in SI engine and action to be taken to reduce knock is listed below

Variable	Major effect on unburned charge	Action to be taken to reduce knock	Can operator Control
Compression ratio	Temp and P	Reduce	No
Mass Inducted	P	Reduce	Yes
Inlet temperature	Temp	Reduce	In some cases
Variable	Major effect on unburned charge	Action to be taken to reduce knock	Can operator control
Chamber	Temp	Reduce	Not ordinarily
Spark timing	Temp and P	Retard	In some cases
A/F : Ratio	Temp and Time	Make very rich	In some cases
Turbulence	Time	Increase	Somewhat (Through Engine Spend)
Engine speed	Time	Increase	Yes
Distance of flame	Time	Reduce	No
Fuel (Octane No)	Self- ignition temp	High	No
Inlet air pressure	Temp and density	Reduce	In some cases
Humidity	Reaction time	Increase	No

Temp : Temperature P : Pressure

4.10 Petrol Injection Systems

Modern carburetors, though highly developed, have certain drawbacks as explained below:

1. In multi-cylinder engines the mixture supplied to various cylinders varies in quality and quantity since the induction passages are of unequal lengths and offer different resistances to mixture flow. The mixture-proportion is also affected due to fuel condensation in induction manifold.

2. Carburetors, with their chock tubes, jets, throttle valves, inlet pipe bends, etc., do not give a free flow passage for the mixture. Thus there is loss of volumetric efficiency on this account.

3. The carburetor has many wearing parts. After wear it operates less efficiently.

4. Freezing may take place at low temperatures, unless special means are provided to obviate this.

5. Surging of the fuel takes place when the carburetor is tilted or during acrobatics in aircraft, unless special means are adopted to avoid this.

6. Backfiring may take place and there is risk of fuel igniting outside the carburetor unless flame traps are provided.

Some of the above disadvantages of a carburetor may be avoided by introducing the fuel by injection rather than by carburetion. In petrol injection timing is by no means critical and the fuel may be injected

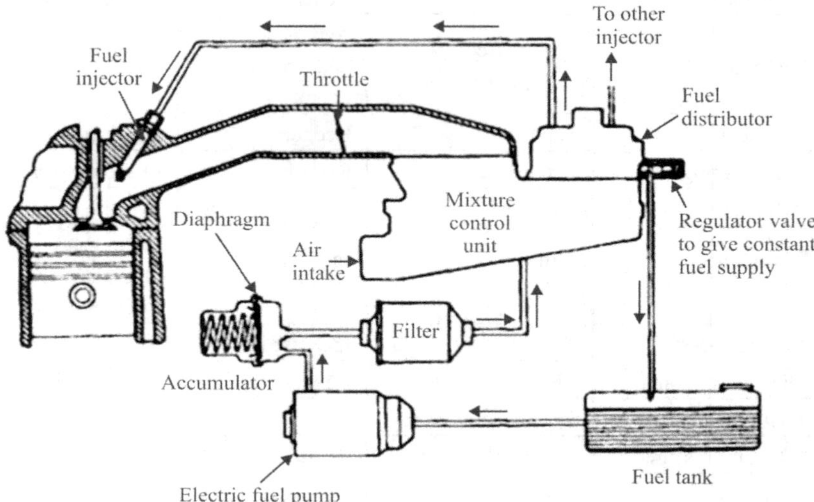

Fig. 4.7 Petrol Injection Systems

during the suction stroke in the inlet manifold at low pressures. The two important methods of injection are continuous injection and timed injections. The schematic of petrol injection is shown in Fig. 4.7.

Types of Injection System

Some of the recent automotive engines are equipped with the fuel injection system, instead of a carburetor for one or more of the following reasons.

1. To improve fuel distribution in a multi cylinder engine
2. To improve air capacity i.e., volumetric efficiency
3. To reduce or eliminate detonation.
4. To prevent loss of fuel during scavenging

The fuel injection system can also be classified as

1. Direct injection
2. Port injection (a) timed, (b) continuous
3. Manifold injection

   The above fuel injection systems can be grouped under two headings, single point injection and multi point injection.

Single Point Injection System: In this system one or two injectors are placed directly into the cylinder. Fuel is sprayed into one point or location at the center of the engine intake manifold. The system is very simple and cheaper. But it creates problem in securing equal distribution of fuel to all cylinders and the fuel particles have a tendency to deposit themselves on the walls of the manifold point injection. This system is sometimes called Throttle Body Injection (TBI)

Multi Point Injection: It has one injector for each engine cylinder. Fuel is injected in more than one location. This is the most common type and is often called port injection.

(a) *Continuous Injection system:* In this system, fuel is sprayed at low pressure continuously into the air supply. The amount of fuel is governed by the air throttle opening, increasing as the throttle is opened. No timing device is used.

   In the case of supercharged engines the fuel may be injected in the form of multiple spray, into the suction side of the centrifugal compressor.

   The continuous injection method has the advantage of promoting efficient atomization of the fuel and uniform mixture strength to the entire cylinder. Also the evaporative effect of the fuel cools the compressed charge and gives a higher volumetric efficiency. This method requires only one fuel injection pump and one injector.

(b) Timed Injection System: The timed injection system is similar to the system used in high speed diesel engines.

    (i) *Multiple plunger jerk pump system*: It employs a pump having a separate plunger for each cylinder and usually a high injection nozzle pressure. The pressure is usually 100 bar to 300 bar. A measured quantity of fuel is delivered into each cylinder during the induction stroke, but at a definite time and over a definite period in this stroke.

    (ii) *Low pressure single pump and distributor system*: This system employs a single plunger or gear pump delivering fuel at relatively low pressure to a rotating distributor which supplies each cylinder with fuel, in turn at the correct time. The pressure is from about 3.5 to 7 bar. The main constituents of a petrol injection system are; fuel feed or lift pump, fuel pump and distributor unit, fuel injection nozzles and mixture controls.

    The mixture controls are automatic for (i) cold starting, (ii) idling, (iii) acceleration,   (iv) normal operation, and (v) full power output.

(c) Multiple Point Injection: The injectors are located immediately before the inlet valve of each cylinder (i.e., one injector for every cylinder is used). Thus the fuel has neither the opportunity to deposit on the walls of manifold, when cold nor the chances of non-uniform distribution in multi-cylinder engines. The relatively high temperature of the inlet valve head helps in vaporization of fuel particles. The principal disadvantage of this system is that the injectors have to be accommodated in the cylinder head and they have to withstand the high temperature in the cylinders and be resistant to combustion deposits.

    (i) *Mechanical Injection*: In this a governor was employed to control fuel supply and a fuel distributor was used to send the fuel to the correct injector. But this system is obsolete now. An electrically driven fuel pressure pump is mounted near the fuel tank. It pumps the fuel at a specified pressure (about 700 Kpa) into a metering distributor. A relief valve returns the excess fuel to the tank, thereby keeping the fuel supply to the metering distributor at constant pressure. The metering distributor supplies fuel to each injector in turn. The quantity of fuel delivered is also controlled in the distributor by engine manifold pressure. The injector is ordinarily held closed by spring until the fuel pressure opens it to deliver atomized spray of fuel. A manual control on the dashboard controls the metering distributor and thereby the quantity of the fuel delivered by it.

(ii) *Electronic Injection*: An electrically driven pump draws the fuel from the tank through a filter and supplied the same to the injectors at a pressure which is held constant by means of a fuel-pressure regulator. The pump draws more fuel than the required and the excess fuel is returned to the tank by the fuel pressure regulator. In this way vapour lock is prevented in the fuel lines. The injectors are held closed by means of spring and are opened by means of solenoids energized by the control signal from the electronic control unit (ECU) which consists of a small preprogrammed analog computer that translates sensor signals into command signals. The strength of ECU signal, which determines the open time of the injector to control the amount of fuel injected, depends upon the engine requirements determined by the ECU from the sensor signals.

Fig 4.8 shows the working principle of a mechanical multi point port fuel injection system Bosch K jetronic system. This system injects fuel continuously in front of the intake valves with the spray directed toward the valves. Thus, about three quarters of the fuel required for any engine cycle is stored temporarily in front of the intake valve and one quarter enters the cylinder directly during the intake process. Air is drawn through the air filter flows past the airflow sensor, past the throttle valve fuel, not required by the engine goes back to the tank. The most important part of the system, the mixture is into the intake manifold and into the individual cylinders. A roller cell pump sucks the fuel out of the tank and the fuel passes through the fuel accumulator filter and fuel distributor. There is a primary pressure regulator in the fuel distributor for maintaining the fuel pressure constant. Excess fuel not required by the engine goes back to the tank.

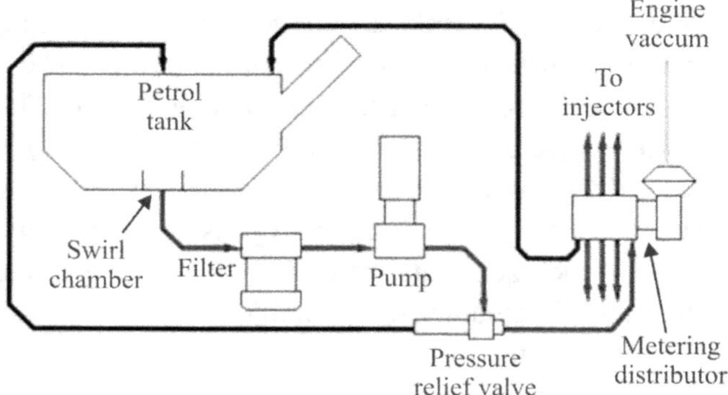

Fig. 4.8 Operation of Mechanical Petrol Injection System

The most important part of the system is the mixture control unit consisting of air flow sensor and the fuel distributor. It provides the desired metering slits in the fuel distributor. The differential pressure valve placed in downstream of each. These metering slits maintain a constant pressure drop across the slits for different flow rates.

Fig 4.9 shows a schematic diagram for electronic multipoint fuel injection system with air flow meter Bosch Jetronic system. The electronic control injection system has fewer mechanical components liable to wear and therefore, more accurate control is possible for wider range of variables. The air flow meter placed upstream of the throttle, measures the force exerted on a plate being displaced by the flowing air stream and provides a voltage proportional to the airflow rate. The injectors are solenoid actuated that make lower delivery pressure possible and delivery can be made through all valves simultaneously. Thus, the injection system is one simpler. In order to ensure uniform fuel distribution to all cylinders, half the required amount of fuel is injected into each part twice, over two separate intervals during each four-stroke cycle.

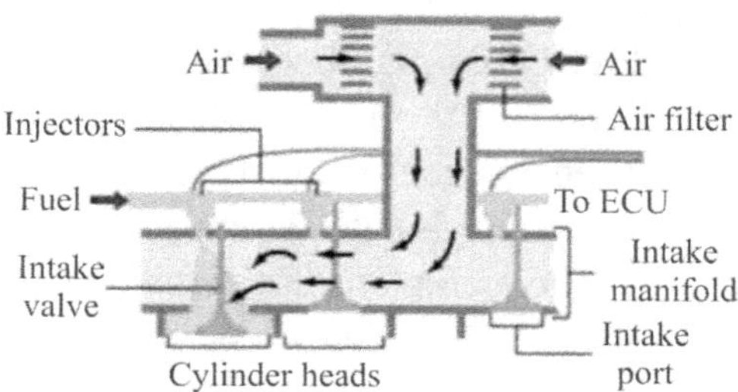

Fig. 4.9 Schematic Diagram of Electronic Petrol Injection System

Advantages

1. Higher power
2. Increased volumetric and thermal efficiency
3. Low specific fuel consumption
4. Simplified induction manifold free from strangled entry
5. No necessity to maintain a stable ignition mixture
6. No necessity of induction heating
7. Free from icing trouble
8. Avoidance of acceleration disturbances during cornering and braking
9. Quick starting and warm-up
10. Greater freedom in choice of fuel
11. Lower exhaust emission
12. Higher quality fuel distribution

Disadvantages

1. Higher initial cost
2. Higher maintenance cost
3. Complicated design
4. Difficult in operation
5. Difficult in servicing

Review Questions

1. Explain briefly the air-fuel mixture requirements of petrol engine from no load to full load.
2. Write a short note on Feed back control carburetor
3   What do you meant by combustion? List and explain the three stages of combustion in SI engine.
4. Discuss the effect of various Engine variables on Ignition lag?
5   Discuss the effects of engine variables on flame propagation?
6   Discuss the basic requirements of a SI engine combustion chamber? What are the various types of combustion chamber used in SI engines? Explain them briefly.
7. Explain the phenomenon of knock in SI engines. What are its harmful effects?

8.   What action can be taken with regard to the various engine variables in order to reduce the possibility of knocking in SI engine? Why?

9.   Explain the effect of various engine variables on SI engine knock (Factors affecting knocking).

10.   What is petrol injection? What are its advantages and disadvantages? Describe with sketch any petrol injection system.

CHAPTER 5

Combustion in CI Engines

- Stages of combustion
- Fuel - Injection systems
    - (i) Direct injection
    - (ii) Indirect injection
- Combustion Chambers
- Fuel spray behaviour, spray structure, spray penetration and evaporation
- Air motion
- Turbo charging

5.1 Stages of Combustion of CI Engine

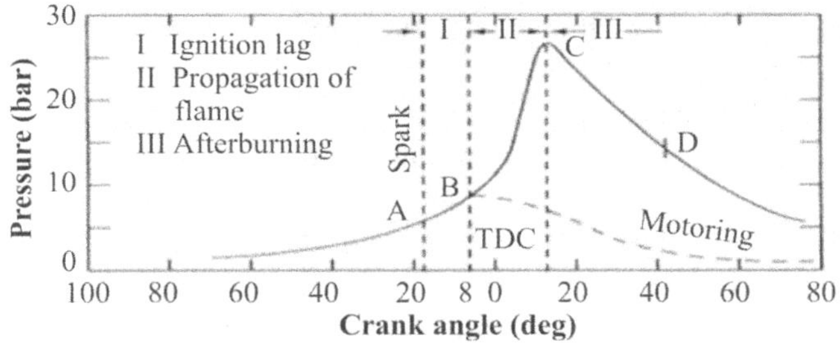

Fig. 5.1 Time (Degree of Crankshaft Rotation)

The combustion in CI engine is considered to be taking place in four stages as shown in Fig. 5.1.

Stages of combustion of CI Engine

1. **First Stage**

    Ignition delay period: It is the time period between the starting of injection and starting of combustion. During this period some fuel has been admitted but has yet been ignited. The ignition delay is counted from the start of injection to the point where the P-θ Curve separates from the pure air compression curve. The delay period is sort of preparatory phase. If the delay period is more, more fuel will be injected inside cylinder and more will be the pressure rise. This causes diesel knock.

2. **Second Stage**

    Period of rapid combustion: Premixed Flame following ignition. In this second stage the pressure rise is rapid because during the delay period the fuel droplet have had time to spread themselves over a wide area they have fresh air all around them. The period of rapid or uncontrolled combustion is counted from the end of delay period to the point of maximum pressure on the indicator diagram. About one third of the heat is evolved during this period.

    The rate of pressure rise in this stage is depend on

    1. Amount of fuel sprayed in the delay period.
    2. Degree of turbulence.
    3. Fineness of fuel-spray.

3. **Third Stage**

    Period of controlled combustion (probable diffusion flame): The second stage of rapid or uncontrolled combustion is followed by the third stage, the controlled combustion. At the end of second stage the temperature and pressure are so high that the fuel droplets injected during the last stage burn almost as they enter and further pressure rise can be controlled by purely mechanical means i.e., by the injection rate. The period of controlled combustions assumed to end at maximum cycle temperature. The heat evolved by the end of controlled combustion is about 70 to 80 percent of the total heat of the fuel supplied during the cycle.

To these three stages of combustion, first proposed by Ricardo, a fourth stage can be added late burning or after burning. This stage may not be present in all cases.

4.   Fourth Stage

After burning: Theoretically it is expected that combustion process shall end after the third stage. However, because of poor distribution of the Fuel particles, combustion continues during part of the remainder of the expansion stroke. This after burning can be called the fourth stage of combustion. The duration of the after burning phase may correspond to 70-80 degrees of crank travel from Top Dead Centre (TDC) and the total heat evolved by the end of entire combustion process is 95 to 97% and 3 to 5% of heat goes as un-burnt fuel in exhaust.

5.2   Requirements of a Diesel Injection System

The following requirements should be fulfilled by the diesel injection equipment:

1.   The fuel should be introduced into the combustion chamber within a precisely defined period of the cycle.

2.   The amount of the fuel injected per cycle should be metered very accurately. The clearances between the working parts of a fuel pump as well as the size of the orifice are very small.

3.   The rate of injection should be such that it results in the desired heat release pattern.

4.   The quantities of the fuel metered should vary to meet changing speed and load requirements.

5.   The injected fuel must be broken into very fine droplets, i.e., good atomization should be obtained.

6.   The spray-pattern must be such that it results in rapid mixing of fuel and air.

7.   The beginning and the end of injection should be sharp. i.e., there should not be any dribbling or after-injection.

8.   The injection timing, if desired, should change to suit the engine speed and load requirements.

9.   In the case of multi cylinder engines, the distribution of the metered fuel among various cylinders should be uniform.

To accomplish the objectives of precise metering, distributing, timing and atomizing the following functional elements are required in a fuel injection system.

Diesel Engines are divided in to two basic categories according to their combustion chamber design.

(i)   Direct-injection (DI) Engines: A single open combustion chamber into which fuel is injected directly.

(ii) Indirect-injection (IDI) Engines: Chamber is divided into two regions and the fuel is injected into the "pre-chamber" which is connected to the main chamber.

Direct-Injection Systems

- The combustion chamber is usually a shallow bowl is the crown of the piston and a general multi-hole injector is used.

- Air swirl in general is achieved by suitable design of inlet port.

- The energy of injected fuel is sufficient to achieve proper mixing of fuel with air.

Indirect-Injection Systems

Two Types of IDI System are:

(i)   Swirl chamber system.

(ii) Pre-chamber system.

   - During compression, air is forced from the main chamber above the piston into the auxiliary chamber, through the nozzles.

   - At the end of compression, a vigorous flow is set up in auxiliary chamber.

   - Fuel is injected at lower injection pressure into the auxiliary chamber.

   - Combustion starts in the auxiliary chamber – pressure rise because of combustion forces fluid into the main chamber where it mixes with air.

   - A glow plug which is heated prior to starting the engine to ensure ignition of fuel is provided in the pre-chamber as a cold starting aid.

5.3   Fuel Spray Behaviour and Spray Penetration

Spray Characteristics

The following general relations apply to different combustion chambers.

1. In open chambers in the absence of strong air motion, the spray must be directed to the various parts of the combustion by a multi-hole nozzle.

2. Divided combustion chambers can usually be made to give satisfactory performance with a single hole nozzle.

3. Strong air motion, or swirl, is always desirable; its presence makes performance less sensitive to spray variation.

4. Spray duration at full load should not exceed about 30 crank degrees.

5. Multiple injections, such as those caused by needled bounce, should be avoided.

The combustion in CI engines is strictly a local phenomenon. It depends upon the local pressure and temperature and upon the mixing of fuel particles with compressed air to support a self-sustaining exothermic chain reaction. The injection of fuel starts a few degrees before the TDC and the fuel is introduced into the combustion chamber through one or more nozzles or orifices with a large pressure difference between the fuel supply line and the cylinder. Standard injectors usually operate with fuel injection pressures between 200 to 1700 atm. At the time of start of injection, the pressure of air inside the cylinder is between 50 to 100 atm, the temperature is around 1000 K and the density varies between 15 to 25 kg/m^3.

Fig. 5.2 Illustrate the structure of a typical diesel engine fuel spray defining its major parameters. The fuel jet usually forms a cone shaped spray at nozzle exit and the size of the droplets depends upon the jet velocity. At low jet velocity the surface tension forces dominate and the droplet size is larger than the jet diameter. The droplet size decreases with increasing jet velocity because of the relative motion between the liquid particles and the surrounding air particles. Usually, the initial jet velocity is greater than 100 m/s so that the liquid jet leaving the nozzle is turbulent, entrains air, spreads and slows down as the mass flow in the spray increases. The outer surfaces of the jet droplets disintegrate into drops of approximately 10 mm diameter and they first evaporate. A liquid core is created and is surrounded by an envelope of fuel vapour-air mixture. The disintegration of liquid droplets takes place over a finite length called the break-up length. The jet velocity is maximum at the axis, the mixing with air is such that the equivalence ratio is highest on the centerline (fuel-rich along most of the jet) and is lowest (almost zero - unmixed air) at the spray boundary.

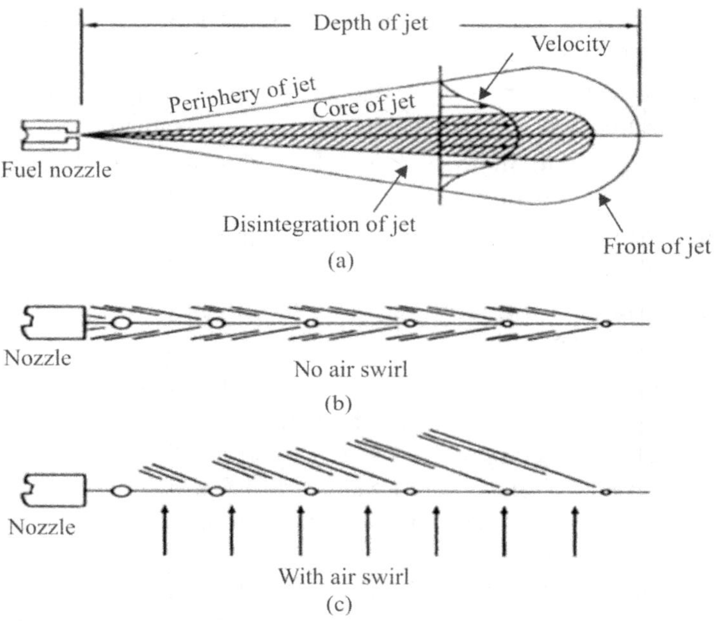

Fig. 5.2 Spray Characteristics of Diesel Engine

The spray tip penetrates into the combustion chamber at a decreasing rate. The speed and extend to which the fuel spray penetrates across the combustion chamber has an important influence on air utilization and fuel-air mixing rates. If the combustion chamber walls are hot and high air swirl is present, it is desirable that the fuel spray impinges on the wall. If the walls are cold and there is little or no air swirl, the mixing rate will be lower and the exhaust emitted will contain unburned and partially burned species. If the spray does not impinge on the combustion chamber into contact with air near the walls of the chamber and there would be poor air utilization. Therefore, the physical characteristics of fuel spray would have an enormous influence on the pressure-crank diagram. And, thus, the injection mechanism along with the mixing of liquid jet with compressed air inside the cylinder has been a subject of extensive research.

Spray Formation

When the fuel is forced through the nozzle holes under high pressure it is disintegrated into fine droplets due to aerodynamic resistance of the dense air present in the combustion chamber. The combustion chamber pressure at the time of injection is about 25 to 35 bar and density 12 to 14 times that of ambient air.

The disintegration of the fuel into fine droplets depends upon the relative velocity of the fuel and on physical characteristics of both fuel and air. The spray angle depends on the density of the medium into which the fuel is sprayed. With high length diameter ratio (L/d) the outlet velocity from the nozzle is very small. The flow is streamlined. With a small L/d ratio the outlet velocity is very high and is turbulent in nature.

A higher L/d ratio gives larger drop size due to their high momentum results in a greater penetration for a given diameter of the orifice and a given fuel pressure. A L/d ratio of about 5 results in turbulent velocity necessary for atomization and L/d ratio of 2 allows a sufficient metal thickness at the nozzle front to withstand injection pressure. For diesel engines L/d ratio is usually between 2 to 5, higher values being used for very large cylinders.

Large droplets have a higher penetration but smaller droplets are required for quick mixing and evaporation of fuel. The fineness of the atomization is indicated by the mean droplet diameter. Smaller the diameter better is the atomization. Most of the droplets are less than 5 microns in size.

An increase in injection pressure results in smaller droplets and uniformity of atomization. The mean droplet diameter increases if the diameter of orifice is increased. An increase in viscosity also increases the mean droplet diameter.

Spray Penetration

The main factors which determine the penetration of the spray are the momentum of the droplets (diameter x velocity) and the density of air in the combustion chamber. Higher the momentum greater is the penetration.

Spray Direction

The relative direction of the fuel spray with air is quite important. The first drops coming into the combustion chamber take heat from the end of the injected spray.

If the direction of spray is same as that of air, the products of first part of combustions are swept from the fuel injected in latter part of injection process. If the fuel is injected upstream of air due to very high relative velocity between air and fuel a very good atomization, and hence, reduced delay is obtained. However, the products of combustion now engulf the newly-arrived droplets and starve them of oxygen necessary for burning. This results in highly smoky exhaust and poor efficiency.

With upstream injection, lower injection pressure and larger orifice diameter can be used. It also results in good starting and part load performance. However, for chambers where the air movement is rather slow upstream injection is not suitable and down-stream injection is invariably used.

5.4 Combustion Chamber Design in CI Engines

In CI engine, combustion occurs because of the high temperature of the compressed air, since that a fuel is ignited with the high temperature of compressed air, it is called auto ignition. The fuel which is in atomized form is slightly colder than the hot compressed air in the cylinder. An appreciable time elapses before the actual combustion starts. This elapsed time is called delay period or ignition lag.

Types of combustion chambers in CI engines

Several types of combustion chambers are used in CI engines as shown in Fig.5.3.

1. The Non-turbulent type (or) open combustion chamber (or) direct injection.

2. The turbulent type (or) divided combustion chamber (or) indirect injection
   (i)   Turbulent combustion chamber
   (ii) Pre-combustion chamber

3. Energy cell and air cell

1. **Non-turbulent type (Open combustion chamber):** This chamber has no turbulence. The upper portion of the cylinder acts as the combustion chamber. High-pressure injectors with orifice nozzles are required to inject fuel properly into compressed air for getting

combustion. This type depends little on turbulence to perform mixing.

2. Turbulent type (Divided combustion chamber):

    (i) *Turbulent combustion chamber:* This combustion chamber depends on turbulence. There is a small antechamber which produces rotary motion to compressed air. Hence the turbulence of air is partially mixed with this air and combustion starts. The pressure built up in the antechamber forces the burning and unburnt air fuel mixture, back into main chamber again increasing the turbulence and producing good combustion.

   (ii) *Pre-combustion chamber:* The piston forces part of the compressed air into the pre-combustion chamber during upward motion. Fuel is injected into the air in the pre-combustion chamber and combustion commences here. The combustion of the fuel and air produces high pressures in the pre-combustion chamber. This creates high turbulence. Thus good combustion occurs.

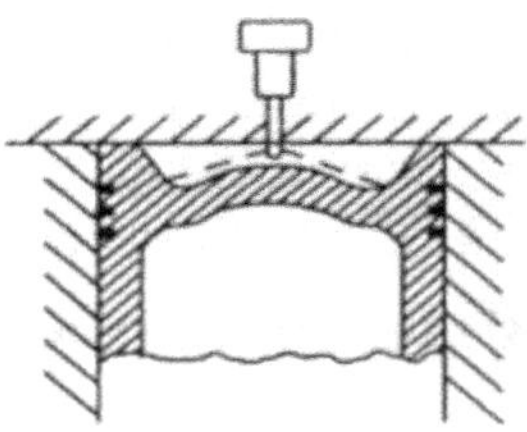

(a) Shallow depth chamber

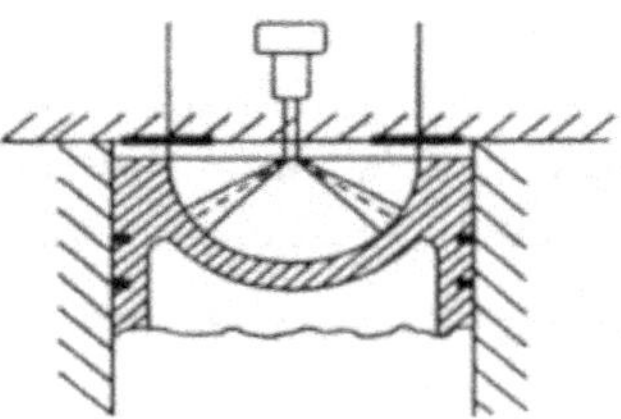

(b) Hemispherical chamber

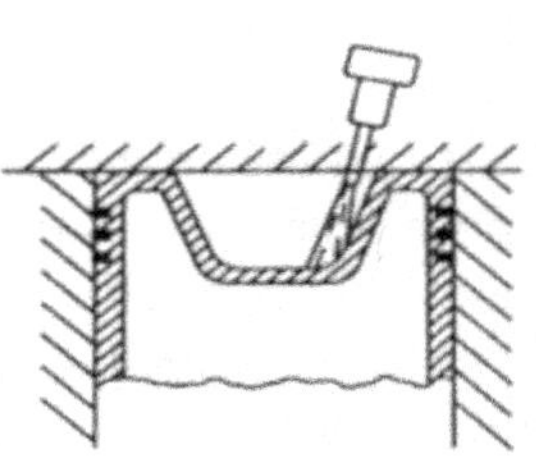

(c) Cylindrical chamber

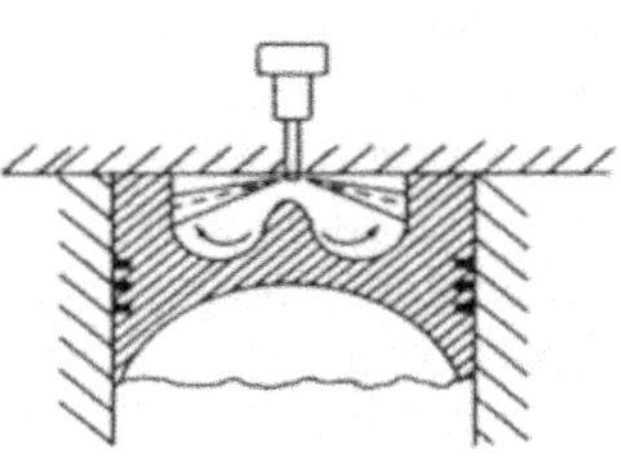

(d) Toroidal chamber

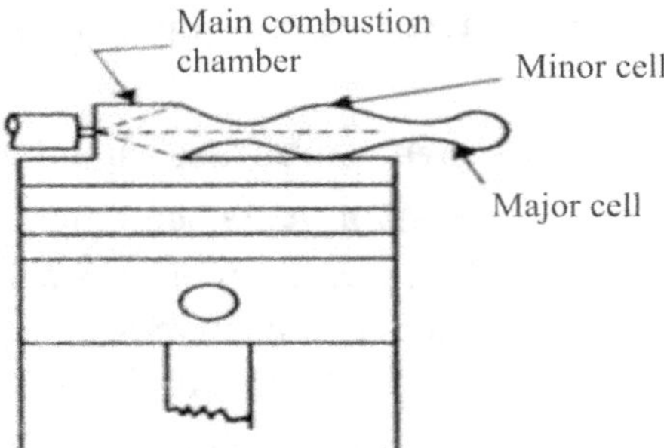

Fig. 5.3 Diagrammatic Illustration of some commonly used CI
Engine Combustion Chambers

2. **Energy cell and air cell:** Energy cell is of turbulent type. When the
    piston moves up part of air is forced into major and minor chambers
    of energy cell. When the fuel is injected by pintle type nozzle, part of
    fuel passes across the main chamber and enters the minor cell. Here it
    is mixed with the entering air. Combustion first commences in main
    chamber. Minor cell and major cell are used to give good turbulence.
    Air cell is the anti-chambers and is used to supply air and secondary
    turbulence to finish the combustion process. The primary turbulence is
    created by providing suitable shape of the piston crown. Combustion
    starts in the main chambers and the air and fuel vapour is forced into
    the air cell. The combustion process is completed in the main chamber
    and air cell is used to provide turbulence inside the main chamber.

Combustion chamber Design for compression ignition Engines

The combustion chamber characteristics have to satisfy the following
objectives in view of the mechanism of combustion in compression
ignition engines.

(i) Mixing of fuel and air take place in the combustion space in
     compression ignition engines, when the fuel is injected unlike the
     spark ignition engine where the same occurs outside in the
     carburetor. Thus combustion chamber must provide adequate
     mixing of fuel and air i.e., short physical delay period. Also the
     air in order to find its fuel should have controlled and organized
     movement termed as Air Swirl and also high relative velocity
     past the fuel droplets.

(ii) Combustion process in Diesel engines should be controlled to
      avoid very high maximum cylinder pressures.

(iii) Also combustion process in Diesel engines should be controlled to avoid excessive rates of pressure rise per degree of crank angle.

(iv) At the same time the combustion should be rapid enough to burn all the fuel very early in the expansion stroke.

Open combustion	Divided combustion chamber
1. Can consume fuels of good ignition quality i.e. of shorter delay and higher cetane number	1. Can consume fuels of poor ignition delay or ignition quality i.e., larger ignition delay or lower cetane number.
2. Requires multiple hole injection nozzles for proper mixing of fuel and air, and also higher injection pressures.	2. It is able to use single hole injection pressures. It can tolerate greater degree of nozzle of fouling.
3. Sensitive to fuel spray characteristics.	3. Insensitive to fuel spray characteristics.
4. Mixing of fuel and air is not so efficient	4. Proper mixing and high air utilization factor.
5. Cylinder construction is simple.	5. More expensive cylinder construction.
6. Easy cold starting	6. Difficult cold starting.
7. Thermally more efficient.	7. Not as efficient because of pressure and heat loss.

5.5 Air Motion in CI Engines: Squish, Swirl, Directional Movement and Turbulence

Air motion

The design of C.I. Engine chamber is more important and difficult than S.I. engine. This is because the mixing of fuel, evaporation and burning takes place inside the combustion chamber. As the evaporation of fuel is carried out inside the combustion chamber, high turbulence of air must be created inside the combustion. This is only possible if the inlet air is provided a swirling motion. To assist the mixing process some turbulence or swirl is deliberately created by careful design of inlet port and combustion chamber.

Swirl denotes a rotary motion of the gases in the chamber, more or less about the chamber axis.

Turbulence denotes a haphazard motion of the gases.

Squish is name given to the flow of air radially inwards towards the combustion recess by squeezing it out from the space between the piston and cylinder head as they approach each other at the end of compression stroke.

There are mainly three methods of providing swirl to the air used for burning fuel. The swirling air creates high turbulence, provides better mixing and high rate to fuel evaporation. The effect of this is to reduce delay period and provide better thermal efficiency.

1.   **Induction Swirl**

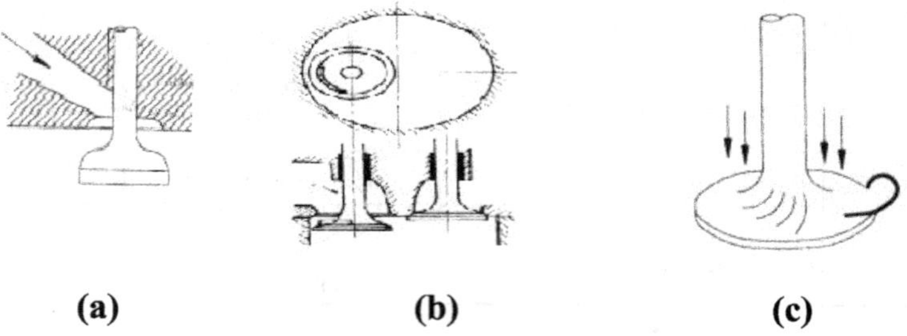

(a)   (b)   (c)

Fig. 5.4 Induction Swirl

This is provided during the suction stroke as the air entry is made tangential or forcing air to take a rotational movement. The induction swirl can be augmented by squish. The induction swirl is used with open type of combustion chamber. Fig.5.4 (a) shows a tangential port method for producing swirl. Fig.5.4 (b) shows the method of producing swirl by making one side of the inlet valve so that air is admitted only around a part of the periphery of the valve and in desired direction. Fig.5.4 (c) shows the method of producing air swirl by casting by lip over one side of the inlet valve.

2. **Compression Swirl or Squish Swirl (Fig. 5.5)**

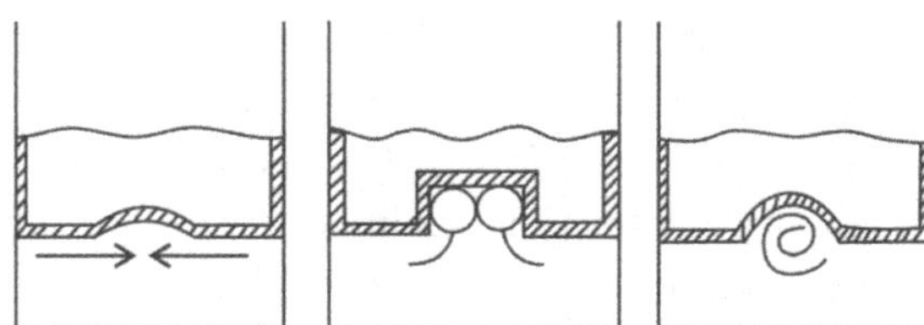

Fig. 5.5 Compression Swirl

Compression swirl is used in divided type of combustion chamber. In this method, Air is forced to the center cavity of the piston during compression stroke. Therefore, this is known as compression swirl. The squishing of air is` created and forced to enter tangentially in the piston cavity when the piston reaches to TDC. The advantages of compression swirl are:

(i) Greater proportion of air retained in the cylinder is consumed and therefore higher m.e.p. is attained.

(ii) Since the injector is at one corner there is freedom to use larger valves with a free entry and so maintain high volumetric efficiency and therefore m.e.p. up to considerable higher speed.

(iii) Due to high swirl, pintle type of nozzle having self-cleaning properties can be used with low injection pressures.

Disadvantage in this design is that combustion products must also pass out through the same restricted passage at still higher velocity and at a time when their temperature and pressure are at maximum; this results in considerable loss of heat to the walls of the passage. This loss can be reduced by using a heat-insulated member.

3. **Combustion Swirl:** This is created due to partial combustion therefore known as combustion inducted swirl. This method is used in pre combustion chamber and air cell chamber. In this design a anti-chamber is connected to the main chamber by a

passage way which is made more restricted than that in a compression swirl combustion chamber. In this method, upward movement of piston during compression forces the air tangentially and the fuel is injected in the pre-combustion chamber. When the combustion of accumulated fuel during delay period burns rapidly in pre-combustion chamber as A/F mixture is rich and forces the gases at a very high velocity in the main combustion chamber. This creates a very high temperature and provides better combustion.

The disadvantages of pre-combustion chamber are:

1. Loss of heat due to high velocity and restricted space for the valves.

2. Cold starting will be difficult as the air loses heat to the chamber walls during compression.

Comparison between induction swirl and compression swirl

Induction Swirl	Compression Swirl
Advantage	**Disadvantage**
1. High excess air (low temperature) low turbulence (less heat loss). Therefore indicated thermal efficiency high.	1. Less excess air, lower indicated efficiency 5 to 8% more fuel consumption. Decreased exhaust valve life.
2. Due to low intensity of swirl easier starting.	2. Cold starting trouble due to high heat loss to strong swirl, greater S/V ratio.
3. No additional work for producing swirl. High mechanical efficiency and hence high brake thermal efficiency.	3. Work absorbed in producing compression swirl, mechanical efficiency lower.
4. Used with low speeds. Therefore low quality of fuel can be used.	4. Cylinder more expensive in construction.
Disadvantages	**Advantage**
1. Weak swirl, multiorifice nozzle, high injection pressure, closing of holes, High maintenance.	1. Single injector, Pintle type (self cleaning) less maintenance.

Contd...

Disadvantages	Advantage
2. Influences minimum quantity of fuel. Complication at high loads and idling.	2. Large valves, high volumetric efficiency.
3. Shoruded valves, small valves, low volumetric efficiency.	3. Greater air utilization due to strong swirl, smaller (chapter) engine.
4. Weak swirl, low air utilization (60%), lower MEP, large size (costly) engine.	4. Swirl proportional to speed, suitable for variable speed operation.
5. Swirl not proportional to speed. Efficiency not maintained in a variable speed engine.	5. Smooth engine operation.

5.6   Factors Affecting Knocking in CI Engine

Diesel knock generally starts at the very beginning of the combustion process due to sudden auto-ignition at different points inside the combustion chamber where as the knocking in the SI engine generally starts near the end of combustion period. Because of this dissimilarity in the time of starting of the knock in SI and CI engines, the conditions, which tend to increase detonation in SI engine, may tend to reduce knocking in CI engines. Therefore good SI fuels are poor CI fuels; this is justified by the following factors. The comparison of knock in SI and CI engine is schematically represented in Fig.5.6.

The various factor affecting knock in CI engine are

  (i)   Compression ratio

  (ii)   Inlet temperature

  (iii)   Self ignition temperature of fuel

  (iv)   Time lag of ignition of fuel and

  (v)   Combustion chamber wall temperature

      (a) *Compression ratio*: If compression ratio is increased then air temperature and pressure at the end of compression also increases thereby decreasing the delay period. So the tendency for detonation increases by increasing the compression ratio in SI engines.

      (b) *Inlet temperatures*: Any increase in the inlet temperature increases the temperature at the end of compression thereby increasing the tendency to knocking in case of SI engines.

   (c) ***Self ignition temperature***: The self-ignition temperature is the temperature of auto ignition of charge. This causes detonation in SI engines, however incase of CI engines, the early auto ignition is necessary to avoid knocking.

   (d) ***Time lag of ignition of fuel***: The time lag of ignition of fuel should be short to avoid knocking in CI engine. However, short ignition time causes knocking in SI engine.

   (e) ***Combustion chamber wall temperature***: If the temperature of the combustion chamber wall is high then auto ignition of charge takes place early causing detonation in the SI engine. On the other hand quick auto ignition due to high wall temperature helps in reducing knocking in CI engine.

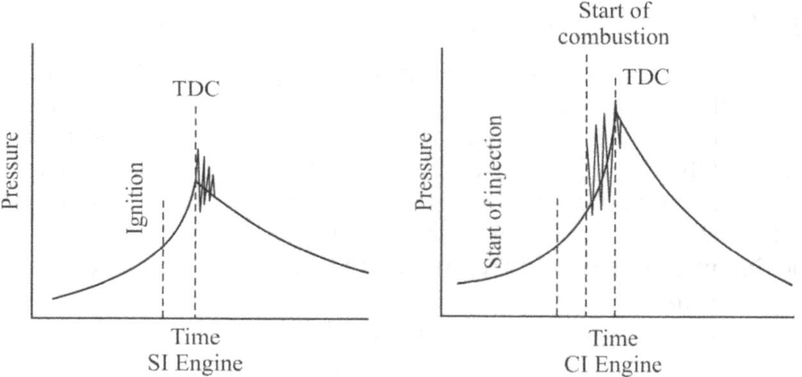

Fig. 5.6 Knock in SI and CI Engine

Characteristics tending to reduce Detonation or knock

S. No.	Characteristics	SI Engines	CI Engines
1.	Ignition temperature of fuel	High	Low
2.	Ignition delay	Long	Short
3.	Compression ratio	Low	High
4.	Inlet temperature	Low	High
5.	Inlet pressure	Low	High
6.	Combustion wall temperature	Low	High
7.	Speed, rpm	High	Low
8.	Cylinder size	Small	Large

5.7   Delay Period and Factors Affecting the Delay Period

The delay period is the time between the start of injection and start of ignition or combustion. The delay period is made of two parts. Physical delay is the time required for the fuel to mix properly with the air as preparation for combustion. The chemical delay is the time from the end of physical delay to the time of ignition. During the physical delay, the fuel particles are atomized, mixed with the air and vaporized. The chemical delay is the physical property of fuel and it depends upon many chemical characteristics of the fuel. The physical delay is small for light fuels, for heavy and viscous fuels, the physical delay is one of the controlling factors. The physical delay can be reduced using high injection pressure and better turbulence. The factors with affect the delay period are discussed below.

1. **Temperatures and Pressure of Intake Air:** The delay period is reduced either with increased temperature or pressure of the intake air because; the vaporization rate increases with either of temperature or pressure or both. The rate of pressure rise during combustion also decreases with an increase in inlet pressure and gives smooth operation of the engine. Supercharged C.I. engine run more smoothly compared with un-supercharged engines.

2. **Compression Ratio:** The delay period decreases with increasing pressure ratio because both pressure and temperature at the end for compression increase with increasing compression ratio. The rate of pressure increases with increasing compression ratio, but peak pressure increases rapidly with increase in compression ratio. A limit is always imposed on the peak pressure during the cycle from design point of view.

3. **Injection Pressure and Injection Timing:** The delay period decreases with increasing pressure (up to a level) of the injected fuel because it provides between atomization penetration and distribution of the fuel. Minimum delay period occurs when the delay period is equally extended on each side of TDC. This is objectionable because it leads to poor thermal efficiency. Early injection of fuel increases the delay period. Injection of fuel about 11.5 °C before TDC gives minimum delay period.

4.   **Engine Load and Speed:** The delay period decreases with increasing load because the operating temperature of the engine increases. The delay period decreases slightly with increasing engine speed. This is better at higher speeds compared to idling speeds.

5.   **Properties of Fuel:** Less viscous fuel give small delay period and fuels with high self-ignition temperatures increase the chemical delay period. The time required to start the combustion after starting the injection depends upon the self-ignition temperature of fuel for the same physical delay.

6.   **Type of Combustion Chamber:** Most of the research work is done to decrease the delay period by changing the shape of the combustion chamber. Presently different types of combustion chambers are in use. In general, specially shaped combustion chambers give shorter delay period compared to open combustion chambers.

7.   **Cetane Number:** The delay period decreases with increasing cetane number of the fuel. The delay period can be reduced from 130 to 80 of crank angle (or 0.002 to 0.0009 sec) with the use of higher cetane number fuel provided all other conditions are same.

5.8   Turbo Charging

About 27% to 38% of heat input to the engine goes as waste along with exhaust gases. This heat available in the exhaust gases is utilized as a useful heat to run a gas turbine, which will supply more air to the engine by driving a compressor. This process is called turbo charging.

So supercharging the engine by utilizing the heat in the exhaust gases is called turbo charging. In turbo charging the super charger is being driven by a gas turbine which uses the energy in the exhaust gases. There is no mechanical linkage between the engine and the supercharger.

The major parts of a turbocharger are:
1.   Turbine wheel
2.   Turbine housing
3.   Turbo shaft
4.   Compressor wheel
5.   Compressor housing
6.   Bearing Housing

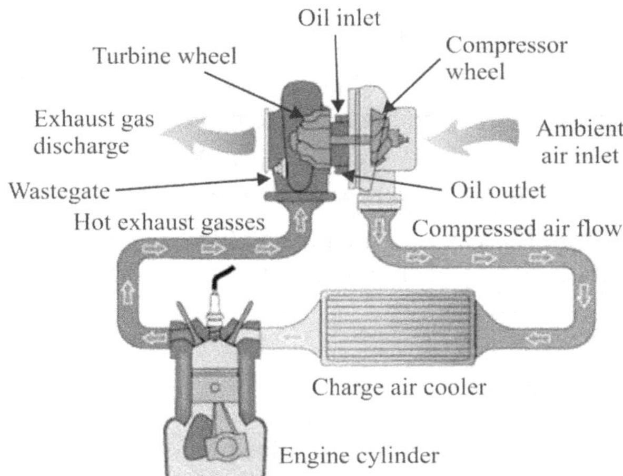

Fig. 5.7 Two Stroke Engine Turbo Charging

The above method of utilizations of heat in the exhaust gases increase the engine power and better thermal efficiency and fuel consumption is achieved. The blow energy available in the exhaust gases, which escape from the engine cylinder at a high velocity, is utilized to a run a gas turbine. This gas turbine is directly coupled to a centrifugal compressor in order to supply extra air to the engine. Centrifugal force throws the air outward. This causes the air to flow out of the turbocharger and into the engine cylinder under pressure. Fig.5.7 shows the method of turbo charging in two stroke engine.

This centrifugal compressor driven by gas turbine is called Turbo Charger. The power developed by the turbocharger is sufficient to run the compressor and overcome its frictional resistance. In case of turbo charging, there is a phenomenon called turbo lag. It refers to the short delay period before the boost or manifold pressure increases. This is due to the time the turbocharger assembly takes the exhaust gases to accelerate the turbine and compressor wheel to speed up. In the turbo charger assembly, there is a control unit called waste gate. This unit limits the maximum boost pressure to prevent detonation in SI engines and the maximum pressure and engine damage. The computer-controlled turbo charging system use engine sensors, a microprocessor and a waste gate solenoid. The solenoid when energized or de-energized by the computer can open or close the waste gate. By this way, the boost pressure can be controlled closely.

Limitations of turbo charging

1. Turbo charging requires special exhaust manifolds.
2. Fuel Injection has to inject more fuel per unit time thus overloading of fuel injection system occurs.
3. After few days of running, the gas turbine blades are subjected to erosion due to the minute solid materials.
4. Since the efficiency of the turbine blades is very sensitive to gas velocity it is very difficult to obtain good efficiency over a wide range of operation.

Review Questions

1. Explain the various stages of combustion of CI engine.
2. What are the functional requirements of a CI engine injection system? How are the injection system classified? Describe them briefly.
3. Explain about Fuel spray behaviour and Spray Penetration.
4. How CI engine combustion chambers are classified? Explain with figures various types of Combustion in CI Engines.
5. Define squish, swirl, directional movement and turbulence. Explain their importance in the design of CI engine combustion chamber.
6. Compare the advantages and disadvantages of induction swirl and compression swirl.
7. The factors that tend to increase detonation in SI engine tend to reduce to knocking in CI engine. Discuses this statement with reference to the various influencing factors.
8. What is meant by ignition delay? Explain the effect of following operating parameter on delay period. (a) Engine Speed (b) Air fuel ratio (c) Injection advance (d) Cetane number (e) Engine Load (f) Injection pressure (g) Compression ratio. What is the importance of delay period?
9. What do you understand by turbo charging? Explain with neat sketch the principle of Exhaust Turbo charging of a single-cylinder engine.

CHAPTER 6

Pollutant Formation and Control

- Pollutants from IC Engines (Sources and Types)
- Formation of NO_x, HC and CO
- Method of controlling Emissions
- Exhaust Gas treatment
    - (i)   Catalytic converter
    - (ii) Particulate trap
- Method of measurement of Exhaust Emissions
    - (i)   Co Infrared absorption
    - (ii) NO_x - Chemiluminescence
    - (iii) Unburned HC - Flame ionization detector
    - (iv) O_2 - Paramagnetic method

6.1 Various Types of Pollutant from IC Engine and their Effects

The various types of pollutant from IC engine and their sources and their harmful effects are listed in Table 6.1.

Table 6.1 Sources of Pollutant in SI Engine

Pollutant	Sources	Harmful effects
1. Carbon Monoxide (CO)	Insufficient supply of air mixture of combustion. Usually 1-2% at Engine exhaust	Increase in mean temperature of earth's surface, Global warming and Green House effect. Affects
		Nervous system, vision, Hearts and Lungs. Restricts function of Hemoglobin to carry O_2 through Blood.
2. Oxides of Nitrogen (NO) Nitric oxide and (NO_2) Nitrogen dioxide	High combustion temperature. Availability of O_2 A/F Ratio 14: 1 and 16: 1. Maximum NO is formed $N_2 + O_2 \rightarrow 2\,NO$; $N_2 + 2H_2\,O \rightarrow 2\,NO + H_2$	Affect Ecological balance, unpleasant odour and give mild irritation in high concentration leads to lung congestion or even death.
3. Unburned Hydrocarbon	Incomplete combustion of HC fuel due to design and operating factors like induction system, combustion chamber design, speed, load, and mode of operation, lean mixture give low HC emission.	They smell bad and in higher concentrations they can be anesthetics. toxic , carcinogenic to humans and attacks the lung tissue
4. Sulphur dioxide	Depends on sulphur content of fuel, sulphur is oxidized to produce sulphur dioxide (SO_2) of which fraction is oxidized to sulphur trioxide (SO_3) combines with water at ambient conditions to form sulphuric acid ($H_2\,SO_4$) and aerosol	Strongly acidic atmosphere, metal corrosion harms living tissues, damage plants, irritation to Eye and Throat Respiratory troubles to children
5. Photo chemical smog	Some HC and Oxides of N_2 react with air in the presence of sun-light to produce photochemical smog.	Damages plant life, reduces visibility, eye irritation and respiratory problem.
6. Soots or particulates	Minute particles found in air may be solid or liquid. Dust soot and fly ash are included in it caused by hydrogenation, polymerization and agglomeration. Participate in smog formation. NO_2 + Particulate + Photo-chemical oxidant + sunlight $\rightarrow$ smog.	• Affects cardiovas-cular system • Heart attacks • Aggravated asthma • Premature death in people with heart or lung disease.

6.2   Source of Pollutants from Petrol Engine

The pollution from large industries, powerhouses and other establishments is taken care of by the skilled engineers by employing various pollution control methods. The most difficult problem is the automobile engine. It is small plant rarely serviced properly. It is operated under various loads and speeds, and is accelerated and decelerated off and on, while being driven on the road. There are millions of these engines at time on the roads. Billions of tons of gasoline and fuel oil is consumed and therefore billions of tons of the products of combustion are discharged in the atmosphere.

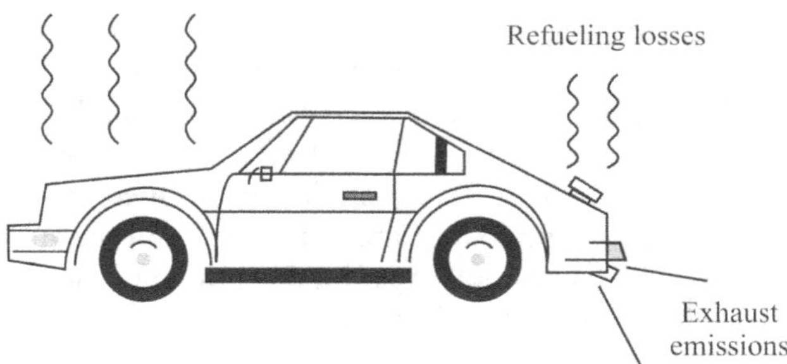

Fig. 6.1 Main Source of Pollutants

The pollutants are discharged from four sources within the combustion engine (Fig. 6.1).

1. **The exhaust pipe (Combustion):** This is the main source, and is about 65% to 85% in Sl engine. It discharges burnt and unburnt hydrocarbon, various oxides of nitrogen NO_x, Carbon monoxide CO and traces of alcohol, aldehydes, ketones, phenols, acids and many others.

2. **Crank case breather:** It discharges about 20% of burnt and unburnt hydrocarbons, because of crankcase blow by.

3. **The fuel tank breather:** Particularly in hot climate there occurs evaporation loss of more volatile hydrocarbons of about 5%.

4. **The carburetor:** Stop and go driving in hot climate leads to evaporation loss. Also very often few fuels get spilled. All this also amounts to about 5% to 10% loss of hydrocarbon.

6.3 Constituents of the Exhaust Emissions from Petrol Engines

Exhaust Emissions: The perfect combustion would give in the exhaust only CO_2 and H_2O (Vapour) and unused air and free N_2. But unfortunately such perfect combustion is never obtainable. Thus we have discussed that many constituents which fall in the category of pollutants are present in the products of combustion. The tailpipe exhausts emission form a major source of emissions in automotive engines. The imperfect combustion gives CO (very poisonous gas and the deadliest pollutant), unburnt hydrocarbon (Various combinations of carbon and hydrogen, and partly burnt hydrocarbons approximated as CH in the results discussed before) in the exhaust emissions. The hydrocarbon also plays an important role in the formation Smog. The Smog is the result of complex sequence of chemical reactions, which include photochemical reaction, which get energy from the ultraviolet end of the solar spectrum. Photochemical smog or automotive smog occurs on hot, dry summer days with high ozone concentration and moderate decrease in visibility. In addition to CO and HC, the different oxides of nitrogen namely NO_x also form pollutant. Some of the oxides are very toxic and harmful. Added to that list is the large organic compound pollutants namely, ketones aldehyde etc., which also help in the formation of Smog, in the presence of sunlight. The petrol, which contains Tetra Ethyl Lead (TEL), to improve the octane number of fuels and suppress knock leads to poisonous compounds of lead. But fortunately, the latest advances in the development of fuels for SI engine, the lead compounds are not present.

6.4 Formation Mechanism of NO_x, HC and CO in Petrol Engine Exhaust

The formation mechanism of pollutants (unburned HC, NO and CO inside the cylinder of a conventional SI engine can be illustrated qualitatively.

Fig. 6.2 shows the four different phases of operation - compression, combustion, expansion, and exhaust. The emissions of unburned HC have several different sources.

(a) When the piston is describing the inward stroke for the compression process, the pressure inside the cylinder increases and some of the gases are forced into crevices (narrow volumes connected to the combustion chamber, the greatest volume lies between the piston, rings and the cylinder walls). These gases are mostly unburned fuel air mixture escaping the primary combustion process, because the entrance to these crevices is too narrow for the flame to enter. The unburned HC comes out of these crevices when the piston describes the expansion and exhausts process and leaves the cylinder.

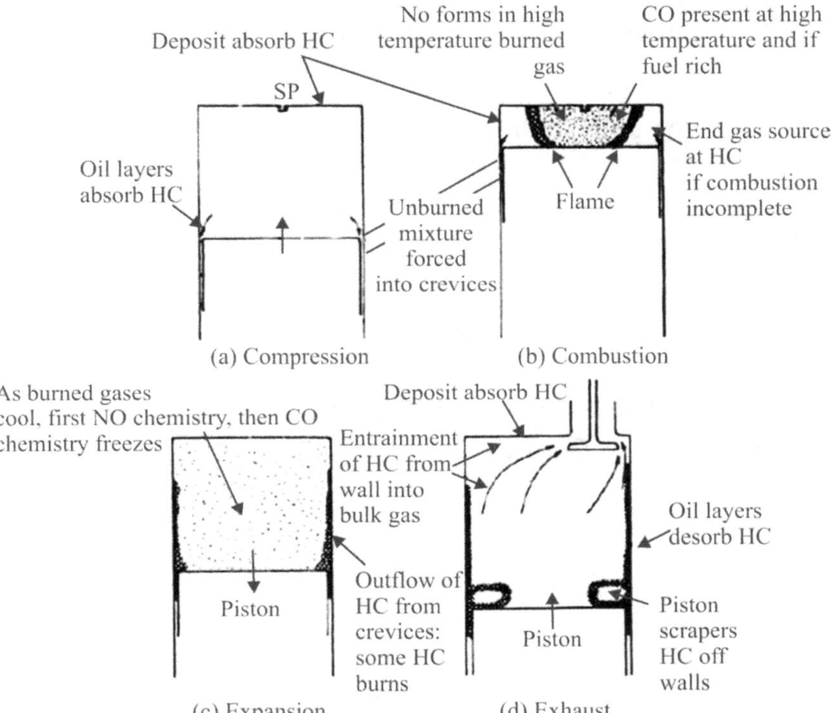

Fig. 6.2 Pollutant Formation Mechanism

(b) The flame gets extinguished when it approaches the wall and a quench layer containing unburned and partially burned fuel air mixture is left on the wall. These deposits also increase the emission.

(c) The engine oil forms a thin film on the cylinder wall, piston and probably on the cylinder head. These oil layers can absorb fuel hydrocarbons before combustion and desorbs them after combustion and a fraction of unburned fuel is permitted to escape through the exhaust.

(d) When the fuel-air mixture is lean or is diluted with increasing amounts of unburned amount of residual gas, the combustion is incomplete and the hydrocarbons leave the engine as exhaust gases. This situation is likely to occur when the air-fuel ratio, spark timing and the fraction of the exhaust recycled for emission controls do not properly match. Nitric Oxide (NO) forms throughout in the high temperature burned gas region behind the flame through chemical reactions involving nitrogen and oxygen molecules, which do not attain chemical equilibrium. The rate of formation of NO is higher in higher temperature of the burned gas. When the temperature of the burned gas decreases during the expansion process, the reactions involving NO freeze and a large concentration of NO leaves through the exhaust.

The formation of CO also takes place during the combustion process. When the fuel-air mixture is rich there is insufficient oxygen to burn all the carbon to CO_2 and the CO_2 dissociates into CO and O at very high temperatures. During the expansion stroke, the temperature of the burned gas falls and the oxidation of CO freezes.

6.5   Effect of A:F Ratio and CO, HC and NO_x Emission from Petrol Engine

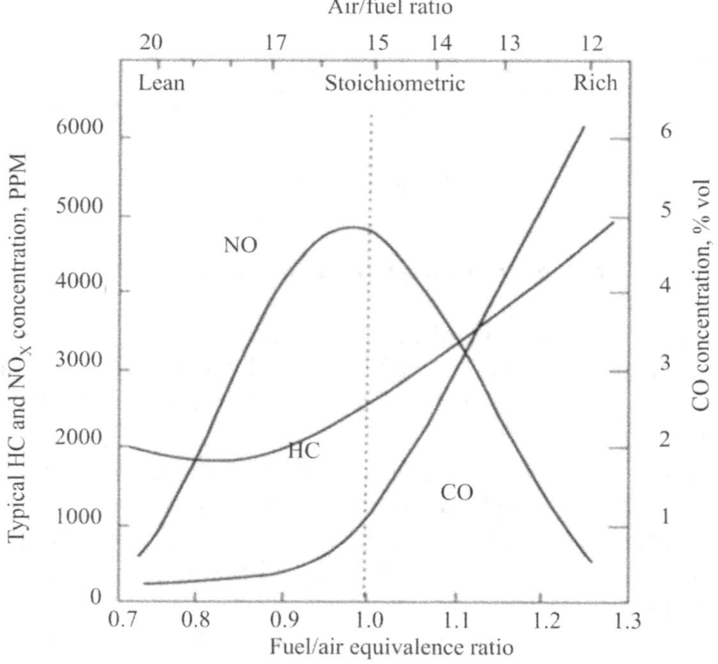

Fig. 6.3 Effect of A/F Ratio on Engine Pollutant

The S.I. engine is always operated at stoichiometric or slightly rich mixture. It is clear from the Fig.6.3 that leaner mixture gives lower emission until combustion quality becomes poor. At the starting of the engine, very rich mixture is supplied, as vaporization is very slow. Thus, until the engine warms up and this enrichment is stopped, CO and HC emission are high at part load condition, lean mixture can be used which will reduce HC and CO emission and moderate NO_x emissions. Use of recycled exhaust to dilute the engine intake mixture lowers the NO_x level but deteriorates combustion quality. Exhaust gas recirculation (EGR) method is used with stoichiometric mixture in many engines to reduce emission.

6.6   SI Engine Emissions Control

Among various methods used for SI engine emission control, the main methods are:

1. Modifications in engine design and operating parameters.
2. Exhaust gas treatment devices.
3. Modification of the fuels to reduce pollutants.
4. Evaporative emission control devices.

1. **Modification in engine design and operating parameters:** For improving the exhaust emission is targeted by the following.

    (i) *Use of lean air fuel ratios:* The carburetor is so designed and adjusted to give lean and stable air fuel ratio of the mixture during idling and cruiser running. The idle speeds may have to be increased to avoid stalling and unstable running of the engine associated with very lean fuel-air ratio. The manifold design often handicaps proper fuel distribution in different cylinders. The manifold design improvements are therefore essential. Fuel distribution is also improved, by inlet air heating raising the temperature of the coolant (often controlled by thermostat) and also use of electronic fuel injection system.

    (ii) *Retard ignition timing:* The ignition timing control is so adjusted as to provide normal required spark advance during cruising and retard the same for idle running. NO_x emissions are reduced due to lowering of maximum combustion temperatures. Also hydrocarbon emission gets reduced due to high exhaust temperatures. But it may be noted that cooling requirements increase. The fuel economy also suffers to some extent accompanied by some power loss. Thus a judicial balance has to be struck between fuel economy, power loss and the pollutants.

(iii) ***Combustion chamber design to reduce wall-quenching areas:*** This is achieved by reducing surface to volume ratio (S/V ratio), reducing squish area.

(iv) ***Lower compression ratio:*** Major change in engine design is to reduce compression ratio from the high > 10 value to the more desirable ratio of about 8 to 8.5. While, it is true that high compression ratio give high thermal efficiencies, such ratio are important for long express highways runs such as those covered by cross-country vehicles and buses. Road tests have shown that compression ratios minor variable for optimum fuel economy in most urban driving patterns. The major variable for good fuel economy are small mass of the vehicle to accelerate with less power, and small frontal area to overcome air resistance with less power. The decrease in compression ratio is becoming very important design parameter. Because it will result in decreased pollutants from the new engine. Besides the octane number requirements will decreases and high leaded gasoline will not be necessary or the lead compound will be phased out to be replaced by un-leaded gasoline.

(v) ***Reduced valve overlap:*** Some of the mixture is allowed to be seeped due to large value overlaps, leading to increase in the level of emissions. This can be controlled by reducing valve overlap. In fact there is always an optimum overlap to take advantage of induction-super-charging in SI engines, and avoid escape of induced mixture.

(vi) ***Modified induction system:*** Special type of carburetors i.e., high velocity carburetors or multi choke carburetors may be used. Metered quantity of fuel may be injected in the manifold. i.e., injection carburetors may be used.

2. **Exhaust gas treatment devices:**

   (i) Exhaust manifold reactor

   (ii) After Burner

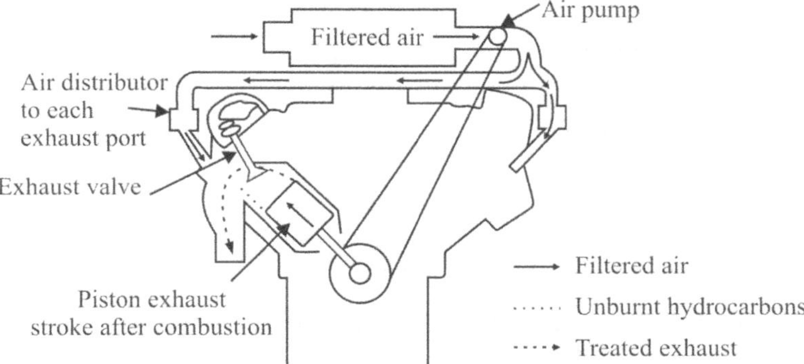

Fig. 6.4 Exhaust Manifold Reactor

(i) ***Exhaust manifold reactor*:** Air injection into the hot escaping exhaust gases reduces emissions of HC, CO and aldehydes by oxidation. The variable controlling the reactions are: Temperature, time, proper mixing of air and exhaust and the composition of the exhaust product. One version of the General Motors Air Injection System is shown in Fig 6.4. Here a positive displacement vane pump, driven by the engine, inducts air from the air cleaner or from separate air filter. The air passed into an internal or external distributing manifold, with tubes feeding a metered amount into the exhaust gases are at high temperature. The injected air reacts with HC, CO and aldehydes to reduce greatly the concentration of such emissions. The injected air is closely metered otherwise it can decrease the exhaust gas temperature.

(ii) ***After burner*:** After burner is nothing but a burner where air is supplied to the exhaust gases and the mixture is burned with the help of ignition system. An addition often after burner to the exhaust system as shown in Fig.6.5 can completely burn the partially burned HC in the exhaust manifold with an intention that the temperature of the exhaust should not fall. After burners have not been successful in curbing the emissions due to the difficultly in sustaining the combustion during low HC emissions.

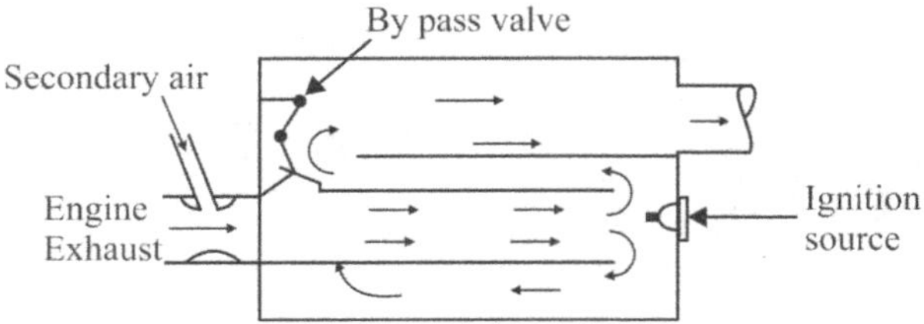

Fig. 6.5 After Burner

3. Modification of the field to reduce pollutants:

Crank case blow by and control: In the sixties the crank case ventilation used to be obtained by a draft tube, which discharged the blow by into atmosphere, and polluting the environments with vapours from the crankcase. This method is no more being used or in fact, it is banned. The positive crank case ventilation method known as PCV is standard and also desirable. The advantages of PCV are

(i)   Atmospheric pollution from blow by is avoided.

(ii)   Ventilation of crankcase gets improved and oil contamination is reduced.

In the PCV system as shown Fig.6.6 Filtered air is drawn from the air cleaner and passes on to the crankcase. The air and the blow by gases pass through flow calibrated PCV valve before being drawn into the intake manifold in order to restrict the air flow while idling when the vacuum is high at the manifold and the blow by is less. PCV valve is spring loaded. At wide-open throttle (WOT) the airflow get unrestricted but flow rate metered by the valve opening.

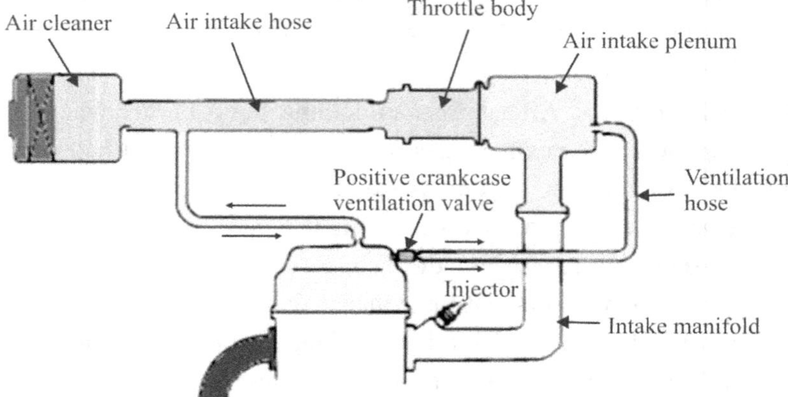

Fig. 6.6 Positive Crank Case Ventilation

The carburetor has to be calibrated, keeping in view the added blow by flow, and since blow by gas and air by-pass the carburetor in entering the manifold, the failure of PCV valve at high vacuum, large quantity of air and also lubricating oil mist will enter the mainfold. This will lead to lean mixtures. Also, the valve getting stuck due to contaminants, will lead to very rich mixtures. It may be noted that the system described does not allow the blow by to go to atmosphere in case of valve failure.

4. **Evaporative Emission Control Devices:** In one of General Motor systems, the filter pipe is sealed with a pressure cap. It comprises a built in vacuum relief to allow to enter as the fuel gets consumed. An inverted bowl is placed inside the tank. It traps air as tank is filled to allow for liquid fuel volume expansion. The air gets released from the bowl by two small orifices in the top. The tank is ventilated to a canister, which has activated carbon particles. This can hold about 200 g of fuel vapor upon shutdown but, when the engine is running, filtered air is drawn through the bottom of canister, removing the intake manifold, in proportion to air flow rate (Fig.6.7).

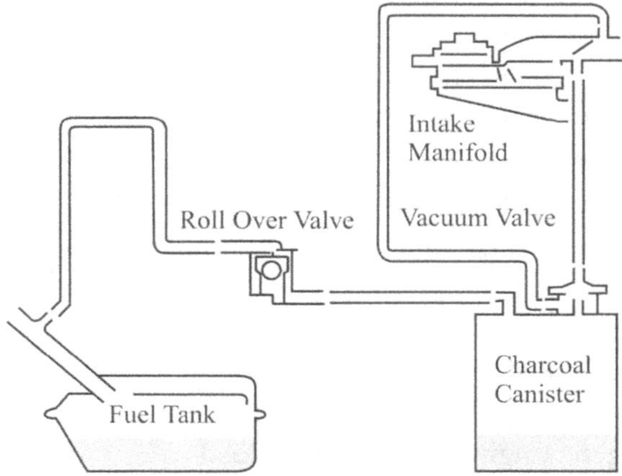

Fig. 6.7 Evaporative Emission Control Device

6.7 Methods to Control Exhaust Emission

Pollution control

The pollution may be controlled by 2 ways

1. Formation of pollutants is prevented as far as possible
2. Pollutants are destroyed after they are formed.

Control of hydrocarbons

Formation of HC may be reduced by following methods

1. Reducing compression ratio
2. Changing design of combustion chamber
3. Changing the design of piston
4. By supplying lean mixture
5. By maintenance of piston and piston ring

Destroying the HC may be done by following methods

1. By supplying air to the inlet manifold
2. By using after burner.
3. By using catalyst converter.

Control of CO

1. By using closed loop control
2. By supplying lean mixture
3. Providing suitable overlap of valves

Methods of destroying CO

1. By using reactor in exhaust manifold
2. By using after burner
3. By using catalytic converter.

Control of oxides of nitrogen

Methods to reduce O_2 are

1. By supplying the exhaust again to inlet manifold
2. By spraying water in the inlet manifold to add moisture to the mixture
3. By using catalyst converter in the exhaust.

Control of smoke and smog

Methods of reducing smoke and smog are

1. Running the Engine with a limited load
2. Maintaining the Engine well.
3. By adding Barium salt in the fuel
4. By using catalyst muffler.

Control of odour

Methods to control odour

1. By using catalyst muffler
2. By changing the injection system in diesel engines.

6.8   Working Principle of a Catalytic Converter

Catalytic converters are used to decrease HC and CO by oxidizing catalyst and NO by reducing catalyst. Thus the factor of temperature, time composition and proper mixing of the exhaust gases are modified by a new variable – the catalyst material. The platinum and palladium act on the HC and CO, while the rhodium affects the NO_x.

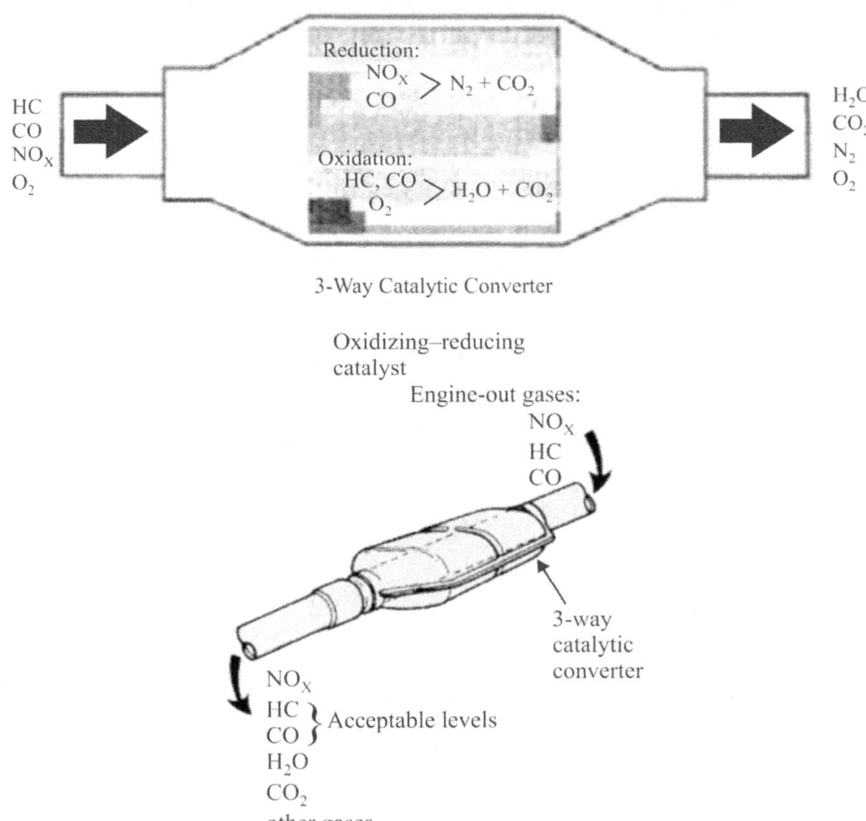

Fig. 6.8 Catalytic Converter

The basic requirements of catalytic converter are

1. It must have low volume heat capacity (kJ/m^3) to reach the operating temperature quickly.

2. Good chemical stability to prevent any deterioration in performance

3. Physical durability with attrition resistance.

4.   Minimum pressure drop during the flow of exhaust gases through the catalyst bed. This will not increase backpressure of the engine.

5.   High surface area of catalysts for better reaction

The practical solution is to pass the exhaust gases through an insulated reactor placed near the engine to ensure quick warm up to temperatures of 300 °C to 600 °C depending upon the catalyst. Inside the reactor, the catalyst is dispersed over large surface area by coating or plating small alumina particles to from a bed (volume) or else upon supports of packed, fine mesh metal screens, properly baffled to ensure mixing as well as time for the chemical reaction. Refer Fig. 6.8 for schematic of a oxidizing and reducing catalysts converter. CO, HC and O_2 from air are catalytically converted to CO_2 and H_2O and numbers of catalysts are known to be effective. Noble metals like platinum and plutonium, and copper, vanadium, iron, cobalt, nickel, chromium etc., are reduction catalyst. Basic concept is to offer the NO molecule an activation site, say on nickel or copper, grids in the presence of CO but not O_2 which will cause oxidation to form N_2 and CO_2. The NO may react with a metal molecule to form an oxide, which then in turn, may react with CO to restore the metal molecule. Hence the name given is catalyst.

The basic operation of the catalyst is to perform the following reaction in the exhaust of the automobile.

Oxidation of CO and HC to CO_2 and H_2

$$CO + H_2O \rightarrow CO_2 + H_2$$

Reduction of NO/NO_2 to N_2

$$2NO + H_2 \rightarrow \tfrac{1}{2} N_2 + H_2O$$

Major drawbacks of catalytic converter are:

1. Cars equipped with such converter should not use leaded fuel as lead destroys complete catalytic activity.

2. Exhaust systems are hotter than normal as a result of exothermic reactions in the catalyst bed.

3. Increase emission of SO_3 if the fuel contains sulphur.

6.9   Constituents of the Exhaust Emissions from CI Engines

The product of combustion from diesel engines also contains essentially the same pollutants as in the petrol engine exhaust, namely HC, CO and for the same reasons. In addition, diesel engines give out other pollutants smoke, odor and noise.

Diesel Smoke and its Control: Smoke is defined as the visible products of combustion. The smoke of a diesel engine is three types - white, blue and black. The white smoke represents liquid particulates or clouds of unburned fuel log. It occurs under cold start, idle or low load and disappears as the load increases. Blue smoke and the particulates are result of burning lubricating oil and fuel additives. It indicates worn piston rings. Black smoke or the soot results from incomplete combustion due to incorrect fuel air ratio and improper mixing. Some methods suggested for control of smoke are:

1. *Derating*: At lower loads the air fuel ratio obtained will be higher and hence the smoke developed will be less as already discussed. However this means a loss of output.

2. *Maintenance*: Maintaining the engine properly, especially the injection system, will not only result in significantly reduced smoke but also keep the performance of the engine at its best. The other methods are changes in combustion chamber geometry fumigation use of smoke suppressant additives. The amount of equipment required to achieve a reduction in smoke which will taper off at higher speeds, at which most of the time the engine will run, do not make it an attractive method of smoke control especially when other methods of smoke control, like use of additives are available. However the strict air pollution regulations can expedite development in this direction.

3. *Smoke suppressant additives*: Some barium compounds, if used in fuel reduce the temperature of combustion, thus avoiding the soot formation, and if formed they break it into fine particles, thus appreciably reducing smoke. However, the use of barium salts increases the deposit formation tendencies of engine and reduces the fuel filter life.

4. *Catalytic mufflers*: Unlike petrol engine the use of catalytic mufflers are not very effective. This is a very small defect on engine smoke. Such devices need much development before they can be used in practice.

5. *Fumigation*: Fumigation is a method of injecting a small amount of fuel in the intake manifold. This helps of pre combustion reactions during compression stroke and reduces the chemical delay because the intermediate products like peroxides and aldehydes react more rapidly with O_2 than hydrocarbons. Reducing

the chemical delay curbs thermal cracking, which is responsible for soot formation.

Diesel noise: In Diesel engine the main exhilarating forces, which set the engine structure into vibration and result in noise are mainly produced from the combustion process. Other sources of noise include the fuel injection system, air intake and exhaust, timing gear and auxiliaries such as cooling fan, supercharge etc. Combustion generated noise can be reduced by shortening the ignition delay period which results in lower rate of pressure rise and lower maximum pressure. The methods are injection, retard modification, turbo charging, increased compression ratio and use of higher octane number fuel

6.10 Working of Infrared Absorption Gas Analyzer for Measuring CO in Exhaust Gases

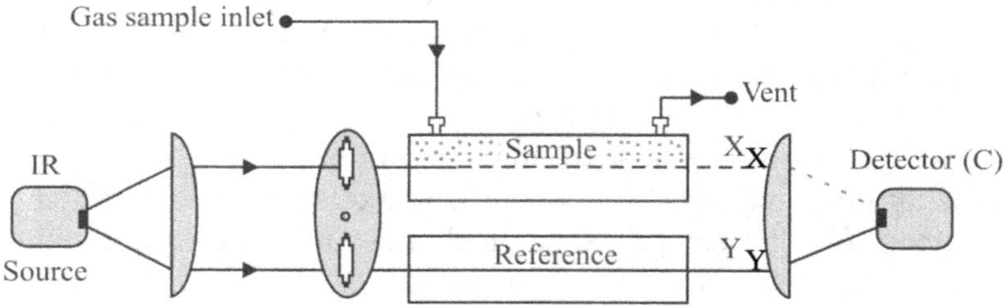

Fig. 6.9 Infrared Absorption (NDIR)

The principle of operation of infrared analyzer is Heteroatomic gases absorb infrared energy at distinct and separated wave length. The absorbed energy also raises the temperature and pressure of the confined gas. Fig 6.9 shows a simplified arrangement. The analyzer has a reference cell which is filled with room air, and a sample cell through which the gas mixture to be analyzed is made to flow. The gas mixture is to be analyzed say for CO_2. Also there is Detector Cell, which is filled with the specific gas to be measured which in this case will be CO_2. The detector cell is divided into two compartments X and Y separated by a diaphragm. This diaphragm is sensitive to pressure changes. In the first instant the sample cell is filled with room air. Since infrared energy absorbed in the two compartments X and Y of the detector cell C will be same, the diaphragm will be stationary and the indicator will read Zero on the scale. After that the sample cell is connected to exhaust gases containing CO_2 and the infrared energy proportional to the concentration of the CO_2 will be absorbed in the sample cell leading to less energy being sent to the sample side of detector. This movement of the diaphragm is amplified electronically and fed to the indicator. This instrument is calibrated mostly again by sending known quantities of CO_2 measured by orsat apparatus.

6.11 Working Principle of Chemiluminescence for Measuring NO in Exhaust Gas

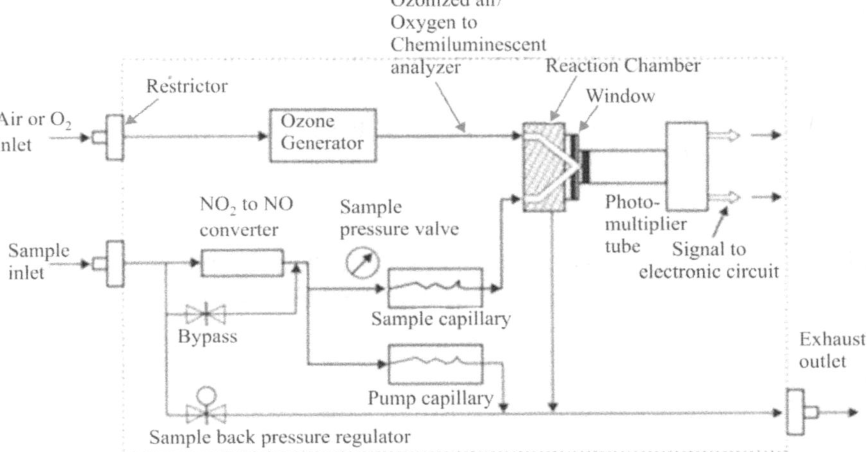

Fig. 6.10 Key Elements in a Chemiluminescence NO_x Analyer

Chemiluminescence

When NO and ozone (O_3) react a small fraction (about 10% at 26.7 °C) of excited NO_2* molecules is produced as per the following reactions:

$$NO + O_3 \rightarrow NO_2 + O_2 \rightarrow NO_2 + O_2 + \text{photon}$$

As the excited molecules of NO_2 * decay to ground state, light in the wavelength region 0.6-3.0 µm is emitted. The quantity of excited NO_2 produced is fixed at a given reaction temperature and the intensity of light produced during decay of excited NO_2 is proportional to the concentration of NO in the sample. This technique depends on the emission of light. Oxides of nitrogen, either NO or NO_x are measured with this technique. The NO in the exhaust gas sample stream is reacted with ozone in a flow reactor chamber. The reaction produces excited NO_2 molecules (NO_2), which in due course can emit light as it converts to its normal state:

Fig. 6.10 shows the arrangement of a NO_x analyzer. The vacuum pump controls the pressure in the reaction chamber and draws ozone and the exhaust sample. The ozone is generated by an electrical discharge in oxygen at low pressure and the flow of ozone is controlled by the oxygen supply pressure and the critical flow orifice (a short length capillary tube). A photomultiplier tube detects the light emitted by the excited NO_2. The signal is then amplified and fed to recorder or indicating equipment. For the measurement of nitrogen oxides (NO_x), NO_2in the sample is first converted to NO by heating in a NO_2- to-NO converter prior to its introduction into the reactor. The instrument can also convert any NO_2 in the sample stream tube by decomposition in a heated stainless steel tube so that the total NO_x concentration can be determined. Since many parameters affect light emission in the reactor, it is essential to calibrate the NO_x analyzer regularly. This is similar to calibrating a NDIR analyzer because calibration gases have to set the zero and check the span.

6.12 Working Principle of Flame Ionization Detector to Measure HC Emissions in the Exhaust

The working principle of flame inonization detector is shown in Fig. 6.11.

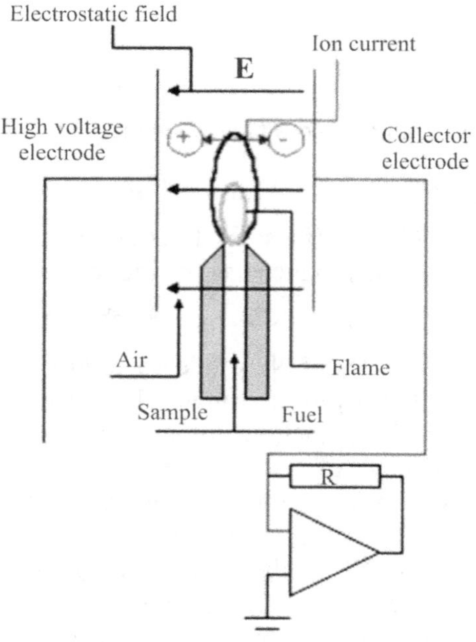

Fig. 6.11 Flame Ionization Detector

When hydrocarbons are burned, electrons and positive ions are formed. If the unburned hydrocarbons are burned in an electric field, then a current flows which is proportional to the number of carbon atoms present. The hydrocarbons present in the exhaust gas sample are burned in a small hydrocarbon air flame producing ions in an amount proportional to the number of carbon atoms burned. The FID is effectively a carbon atom counter. It is calibrated with sample containing known amount of hydrocarbons in the sample line (especially important in diesel exhaust gas), the sample line is often heated. The FID cannot distinguish between different hydrocarbon species. However, gas chromatography can be used for identifying particular hydrocarbon species in the exhaust gas. The gas sample and the carrier gas chromatography can be used for identifying particular hydrocarbon species in the exhaust gas. The gas sample and the carrier gas are passed into the column – a long tube which contains different molecules in discrete groups and the different molecules can be identified by their residence time. The exit of species from the column is detected by measuring the ionization in flame. The chromatograph is calibrated by injecting sample of known gases into the carrier gas.

6.13 Working Principle of Oxygen Analyzer

Oxygen is obviously not an undesirable exhaust emission, but its measurement is very useful in evaluating the air-fuel ratio. Oxygen concentration is usually measured with paramagnetic analyzers.

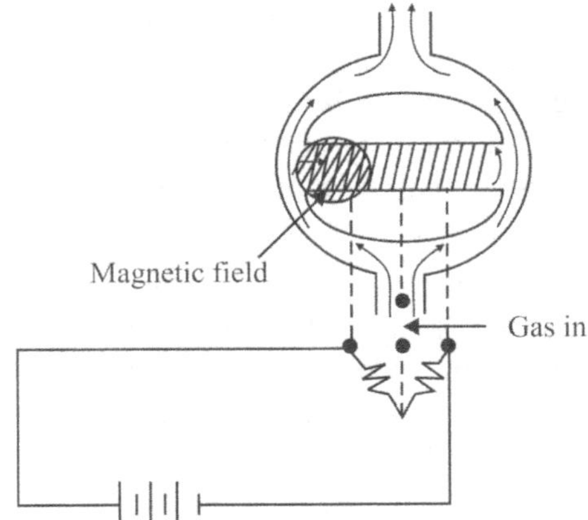

Fig. 6.12 Basis for Paramagnetic Oxygen Analyser

The principle of a thermo magnetic paramagnetic oxygen analyzer is illustrated by Fig.6.12. The heated filament in the cross tube forms a part of Wheatstone bridge. Oxygen is attracted by the magnetic field but when it is heated by a filament, its paramagnetism is reduced and the oxygen flows away from the magnetic field. The flow of the gas cools the filament which changes its resistances and this leads to an unbalance in the bridge. The following precautions should be taken during measurement.

(a) The heated filament will be affected by changes in the transport properties of the gases in the sample.

(b) Hydrocarbons and other combustible gases can react on the filament causing changes in the filament temperature.

(c) The cross-tube must be mounted horizontally to avoid natural convection phenomenon taking place.

6.14 Measurement of Smoke Intensity

Two basic types of smoke meters for measuring smoke density are (i) filter darkening type (ii) Light extinction type.

Smoke meter (light extinction meter): The smoke meter light source consist of incandescent lamp with a color temperature range of 2800 K to 3250 K, or a light source with a spectral peak between 550 to 570 nanometers. The light output is collimated to a beam with a maximum diameter of 1.125 inches and an included angle of divergence within a 6° included angle. The light detector shall be a photocell or photodiode. A collimating tube with apertures equal to the beam diameter is attached to the detector to restrict the viewing angle of the detector to within a 16° included angle. An amplified signal corresponding to the amount of light blocked is recorded continuously on a remote recorder. The light extinction type of meters can measure both white and black smoke while the filter paper type meters can give only black smoke readings. All these smoke meters give reasonable correlations or black smoke.

When light from a source is transmitted through a certain path length of the exhaust gas, smoke *pacity* is the fraction of light that is absorbed in the exhaust gas column and does not reach the light detector of smoke meter. The absolute smoke density is given by the absorption coefficient, k_s which has units of m^{-1} and is given by:

$$k_s = \frac{1}{L}\ln\left(\frac{I}{I_0}\right)$$

where L is length of smoke column in meter through which light from the source is made to pass, I_0 is the intensity of incident light, I is the transmitted light falling on the smokemeter receiver.

Now, mostly light extinction/absorption smokemeters based on Beer-Lambert Law are used. The light extinction type smokemeters are more commonly called as 'opacimeters' as these provide a more realistic measurement of the visible smoke emissions from diesel engines. The light extinction meters can be used for continuous measurements. While the filter type can be used only under steady state conditions.

1. **Hartige smoke meter:** It works on the principle of light extinction. Exhaust sample is passed through tube of about 46 cm length, which has light source at one end, and photocell at the other end. The amount of light passed through this tube is used as indication and meter is calibrated (Fig.6.13).

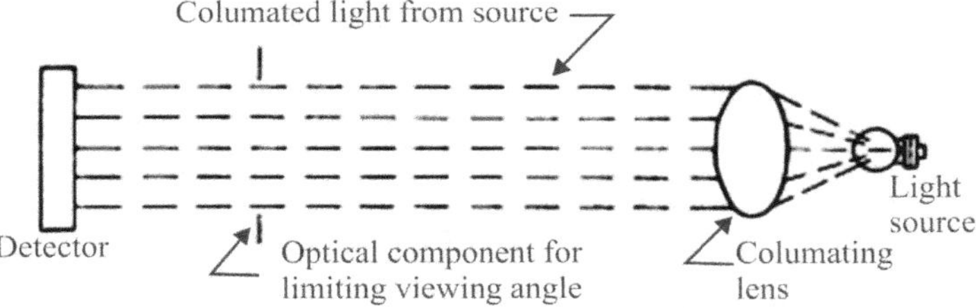

Fig. 6.13 Hartige Smoke Meter

2. **Bosch smoke meter:** It is a filter darkening type of smoke meter. A measure volume of exhaust gas is drawn through a filter paper, which is blackened, to various degrees depending upon the carbon present in the exhaust gas. The density of soot is measured by timing out amount of light reflected from the sooted paper. The specifications of sample volume, filter paper size etc is well defined (Fig.6.14).

3. **Van brand smoke meter:** Van Brand smoke meter is also filter-darkening type. The exhaust sample is passed at a constant rate through a strip of filter paper moving at a pre set speed. A stain is imparted to the paper. The intensity of the stain is measured by the amount of light, which passes through the filter and is indication of the smoke density of exhaust. Bosch and Van Brand smoke meters differ in that the first in the amount of light reflected is the measure of smoke level while in the second the amount of light passing through the filter is used to indicate smoke level.

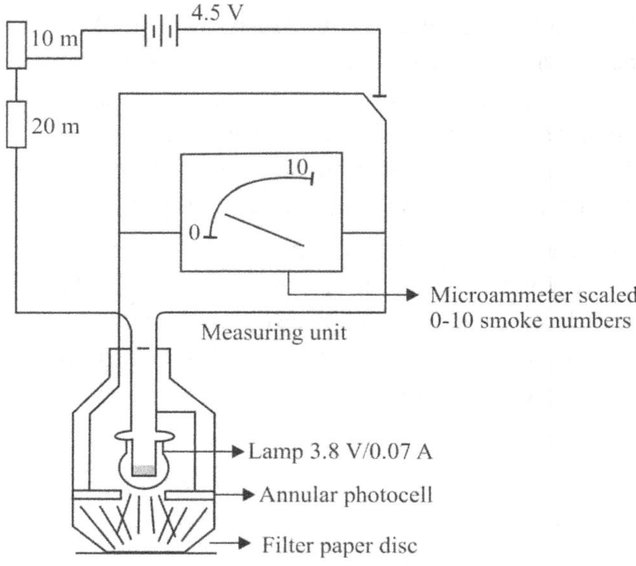

Fig. 6.14 Bosch Smoke Meter

6.15 Particulates Emissions

The exhaust of CI engines contains solid carbon soot particles that are generated in the fuel-rich zones within the cylinder during combustion. These are seen as exhaust smoke and cause an undesirable odorous pollution. Soot particles are clusters of solid carbon spheres. These spheres have diameters from 9 nm to 90 nm (1nm = 10^{-9}). The spheres are solid carbon with HC and traces of other components absorbed on the surface. A single soot particle may contain up to 5000 carbon spheres. Carbon spheres are generated in the combustion chamber in the fuel-rich zones where there is not enough oxygen to convert all carbon to CO_2.

Up to about 25% of the carbon in soot comes from Lubricating oil components, which vaporize and then react during combustion. The rest comes from the fuel and amounts to 0.2%-0.5% of the fuel. Because of the high compression ratios of CI engines, a large expansion occurs during the power stroke. This makes the gases within the cylinder to be cooled to a relatively low temperature. This causes the remaining high boiling point components found in the fuel and lubricating oil to condense in the surface of the carbon soot particles. This absorbed portion of the soot particles is called the soluble organic fraction (SOF), and the amount is highly dependent on cylinder temperature. SOF can be as high as 50% of the total mass of soot. SOF consists mostly of hydrocarbon components with some hydrogen, NO, NO_2, SO_2 and trace amount of calcium, chromium, iron, phosphorus, silicon, sulphur and zinc. Particulate generation can be reduced by engine design and control of operating conditions, but quite often this will create other adverse results. Engine management systems are programmed to minimize CO, HC, NO_x, and particulate emissions by controlling ignition timing, injection pressure, injection timing, and/or valve timing. In most cases exhaust particulate amounts cannot be reduced to acceptable levels solely by engine design and control.

Particulate traps

Compression ignition engine systems are equipped with particulate traps in their exhaust flow to reduce the amount of particulates released to the atmosphere. Traps typically remove 60-90% particulates in the exhaust flow. An exhaust treatment technology that substantially reduces diesel engine particulate emissions is the traps oxidizer. It consists of a temperature-tolerant filter or trap, which removes the particulate material from the exhaust gas. The filter is then cleaned by oxidizing the accumulated particles.

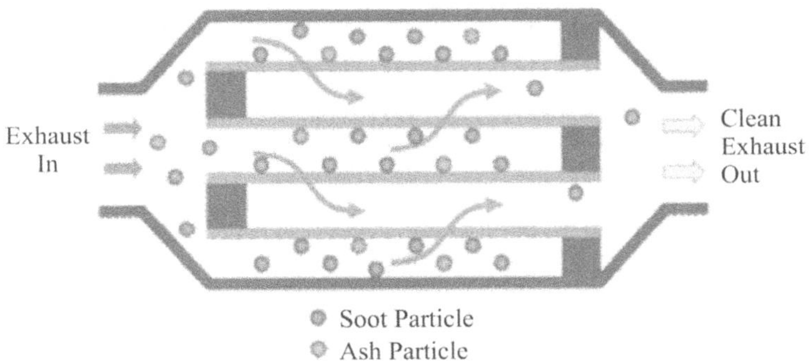

Fig. 6.15 Particulate Trap

The particulate filters include; ceramic monoliths, alumina-coated wire mesh, ceramic foam, ceramic filter mat, woven silica fiber rope wound round a porus tube. Each of these has different inherent pressure loss and filtering efficiency. Diesel particulate matter ignites at about 500 °C to 600 °C. This is above the normal temperature of diesel exhaust and therefore, the exhaust gases flowing through the trap require heating or the ignition temperature by up to 200 °C. Trap oxidizers have been put into use for light-duty automobile diesel engines. Their use with heavy duty diesel engines poses more problems due to higher particulate loading and lower exhaust temperatures (Fig.6.15).

6.16 Emission Standards

Pollutants are known to cause serious health problems. Therefore there are laws on emission standards, which limit the amount of each pollutant in the exhaust gas emitted by an automobile engine. Emission standards have been followed for sometime in the developed countries. India is in favor of the "European Model" developed by the European countries. The European emission norms are called 'Euro Norms'. India has adopted the European emission limits on a slightly modified driving cycle for the light duty vehicles. The overall emission limits and test procedures for the heavy duty vehicles however are the same as those in Europe. In India, the emission limits are being enforced by a time lag of around 5 years as shown in Table 6.2. Euro I emission norms were enforced in the New Delhi Capital Region from June 1999 and Euro II norms have effect from April 2000 throughout India. On October 6, 2003, the National Auto Fuel Policy has been announced, which envisages a phased program for introducing Euro 2-4 emission and fuel regulations by 2010. The implementation schedule of EU emission standards in India is summarized in Table 6.2.

Table 6.2 Emission Standards for Diesel Truck and Bus Engines, g/kWh

Year	Reference	CO	HC	NOx	PM
1992	-	17.3-32.6	2.7-3.7	-	-
1996	-	11.20	2.40	14.4	-
2000	Euro I	4.5	1.1	8.0	0.36*
2005†	Euro II	4.0	1.1	7.0	0.15
2010†	Euro III	2.1	0.66	5.0	0.10
		5.45	0.78	5.0	0.16
2010†+	Euro IV	1.5	0.46	3.5	0.02
		4.0	0.55	3.5	0.03

† Nationwide

†+ National Capital Region (Delhi) and 11 cities Mumbai, Kolkata, Chennai, Bangalore, Hyderabad, Secunderabad, Ahmedabad, Pune, Surat, Kanpur and Agra

6.17 Exhaust Gas Recirculation Method to reduce NO$_x$, Emission

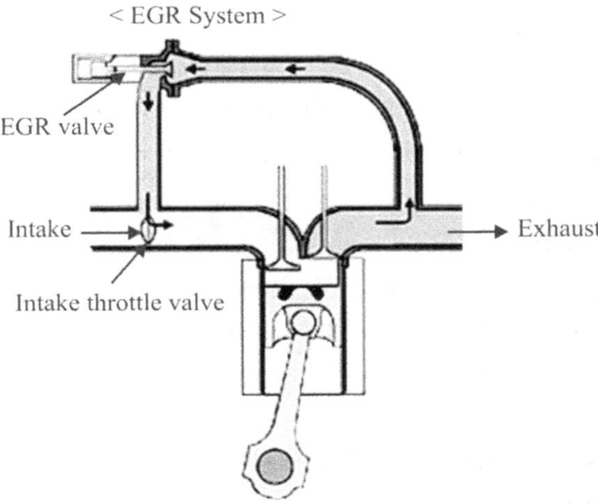

Fig. 6.16 Exhaust Gas Recirculation System

Exhaust gas is partially mixed with intake air to lower combustion temperature for reducing NO$_x$ and fuel consumption. EGR cooler cools down exhaust gas to increase air concentration for complete combustion, reducing Particulate Matter. The most effective way of reducing NO$_x$ emissions is to hold combustion chamber temperatures down. To obtain maximum engine thermal efficiency it should be operated at the highest temperature possible. Exhaust gas recirculation (EGR) is an effective means for control of NO$_x$ emissions. It is achieved by reducing the flame temperature by mixture dilution (Fig.6.16). The simplest and practical method of reducing maximum flame temperature is to dilute the air fuel mixture with a non-reacting parasite gas. This gas absorbs energy during combustion without contributing any energy input. The net result is a lower flame temperature.

Adding any non-reacting neutral gas to the inlet air-fuel mixture reduces flame temperature and NO$_x$ generation; Exhaust gas recycle (EGR) is the one gas that is readily available for engine use. Exhaust gas recycle (EGR) is done by ducting some of the exhaust flow back into the intake system, usually immediately after the throttle. EGR combines with the exhaust residual left in the cylinder from the previous cycle to effectively reduce the maximum combustion temperature. By using EGR to reduce NO$_x$ emissions, a costly price of increase HC emissions and lower thermal efficiency must be paid. Engines with fast burn combustion chambers can tolerate a greater amount of EGR.

Review Questions

1. Give a brief account of air Pollution due to engine

2. What are main source of pollutants from petrol engine?

3. What are the main constituents of the exhaust emissions from petrol engines?

4. Give a brief account of emissions from CI engines. What are the ways of controlling diesel smokes?

5. Explain the formation Mechanism of NO_x, HC and CO in petrol engine exhaust.

6. Discuss the effect of A:F ratio and CO_2, HC and NO_x emission from petrol engine.

7. What engine modifications are used to reduce the emissions in SI engine?

8. Describe with neat sketches used to control the exhaust emissions in SI engine.

9. What is meant by crankcase blow by? How it can be controlled:

10. What are the sources of evaporative emission in petrol engine? How it can be controlled?

11. What are catalytic converters? Explain the principle of working of a catalytic converter

12. With the help of neat diagram, explain the working of infrared absorption gas analyzer for measuring CO in exhaust gases.

13. Explain the working principle of chemiluminescence for measuring NO in exhaust gas.

14. Explain the working principle of flame ionization Detector used to measure HC emissions in the exhaust.

15. Explain the working principle of oxygen analyzer.

16. How smoke intensity can be measured? Describe with sketches the two important types of smoke meters.

17. Write short notes on Emission standards.

18. What do you understand by the term EGR? Explain how EGR reduces NO_x, emission.

19. Describe in detail how particulate emissions are caused. Explain about particulate trap.

CHAPTER 7

Conventional and Alternative Fuels

Conventional and Alternate Fuels: Types of fuels and their characteristics; Introduction, Objectives, Principles of Classification of Fuels, Solid Fuels and their Characteristics, Woods and their Characteristics, Coals and their Characteristics, Manufactured Solid Fuels and their Characteristics, Liquid Fuels and their Characteristics, Petroleum and its Characteristics, Manufactured Liquid Fuels and their Characteristics, Gaseous Fuels and their Characteristics, Natural Gas and its Characteristics, Manufactured Gases and their Characteristics, Alternative fuels; Alcohol, Hydrogen, Natural Gas, Liquefied Petroleum Gas, Properties, Suitability, Engine modifications, Merits and Demerits of above fuels

7.1 Introduction

Fuel is a substance which, when burnt, i.e., on coming in contact and reacting with oxygen or air, produces heat. Thus, the substances classified as fuel must necessarily contain one or several of the combustible elements: carbon, hydrogen, sulphur, etc. In the process of combustion, the chemical energy of fuel is converted into heat energy. To utilize the energy of fuel in most usable form, it is required to transform the fuel from its one state to another, i.e., from solid to liquid or gaseous state, liquid to gaseous state or from its chemical energy to some other form of

176

energy via single or many stages. In this way, the energy of fuels can be utilized more effectively and efficiently for various purposes.

7.2   Principles of Classification of Fuels

Fuels may broadly be classified in two ways, i.e.,

(a)   According to the physical state in which they exist in nature – solid, liquid and gaseous, and

(b)   According to the mode of their procurement – natural and manufactured.

None of these classifications, however, gives an idea of the qualitative or intensive value of the fuels, i.e., their power of developing the thermal intensity or calorimetric temperature under the normal condition of use, i.e., combustion of fuels in mixture with atmospheric air in stoichiometric proportion.

A brief description of natural and manufactured fuels is given in Table 7.1.

Table 7.1 Natural and Manufactured Fuels

Natural Fuels	Manufactured Fuels
Solid Fuels	
Wood	Tanbark, Bagasse, Straw
Coal	Charcoal
Oil shake	Coke
Oil shake	Briquettes
Liquid Fuels	
Petroleum	Oils from distillation of petroleum
Petroleum	Coal tar
Petroleum	Shale-oil
Petroleum	Alcohols, etc.
Gaseous Fuels	
Natural gas	Coal gas
Natural gas	Producer gas
Natural gas	Water gas
Natural gas	Hydrogen
Natural gas	Acetylene
Natural gas	Blast furnace gas
Natural gas	Oil gas

7.3   Solid Fuels and their Characteristics

Solid fuels are mainly classified into two categories, i.e., natural fuels, such as wood, coal, etc., and manufactured fuels, such as charcoal, coke, briquettes, etc.

The various advantages and disadvantages of solid fuels are given below:

Advantages
 (a)   They are easy to transport.
 (b)   They are convenient to store without any risk of spontaneous explosion.
 (c)   Their cost of production is low.
 (d)   They possess moderate ignition temperature.

Disadvantages
 (a)   Their ash content is high.
 (b)   Their large proportion of heat is wasted.
 (c)   They burn with clinker formation.
 (d)   Their combustion operation cannot be controlled easily.
 (e)   Their cost of handling is high.

7.3.1 Woods and their Characteristics

The most commonly used and easily obtainable solid fuel is wood. It is the oldest type of fuel which man had used for centuries after the discovery of the fire itself. In India, wood is used in almost every village, as well as in small towns and cities. In some parts of country such as Kashmir and Mysore, wood is used for industrial purposes as well.

Constituents of Wood

Wood is vegetable tissue of trees and bushes. It consists of mainly cellular tissue and lignin and lesser parts of fat and tar, as well as sugar.

The main constituents of several kinds of wood are given in Table 7.2.

Table 7.2 Constituents of Wood

Type of Wood	Water	Sugar	Fat-tar	Cellular tissue	Lignin
Beach wood	12.57	2.41	0.41	45.57	39.14
Birch wood	12.48	2.65	1.14	55.62	28.21
Fir (Boot)	13.87	1.26	0.97	55.90	26.91
Pine wood	12.87	4.05	1.63	53.27	28.18

The constituents of cellular tissue and lignin of wood are given in Table 7.3.

Table 7.3 Constituents of Cellular Tissue and Lignin of Wood

Constituents	Cellular tissue (%)	Lignin (%)
Carbon	44.4	54.58
Hydrogen	6.2	5.8-6.3
Oxygen	49.4	35.39

The cellular tissue has a definite chemical composition and thus has stable constituents, while those of lignin vary within narrow limits. Hence, the constituent elements of different kinds of wood are slightly variable. Table 7.4 gives the constituents elements of wood and the average values of constituents of wood are given in Table 7.5.

Table 7.4 Constituents Elements of Wood

Element	Pine wood (%)	Birch Wood (%)	Oak Wood (%)
C	50.05	48.45	49.8
H	6.04	5.95	5.81
O + N	43.21	45.26	44.00
Ash	0.70	0.34	0.4

Table 7.5 Average Values of Constituents of Wood

Constituents	Cellular tissue (%)
C	50.00
H	6.00
O	43.10
N	0.30
Ash	0.60

Calorific Value

Engineer A. Marjhevskee determined the calorific values of different kinds of wood with the help of the samples taken out from the same tree at different distances from centre. The calorific values are given in Table 7.6.

Table 7.6 Calorific Values of Woods

Kinds of Wood	Lowest calorific value (cal/kg)	Highest calorific values (cal/kg)
Oak	4729	4750
Birch	4695	4831
Elm	4674	4833
Alder	4745	4839
Pine	4818	5310
Fir	4887	4900
Lrch	4775	4840

Ash

The ash content of wood is negligible. The ash consists of mineral water that is found in the wood itself, with an admixture of some impurities which accrued during transportation, etc. The mineral matter is distributed in the tree rather irregularly. The ash consists of mainly potassium carbonate with varying degrees of calcium, magnesium and sodium carbonate, as well as minute quantities of iron oxides, alumina and silica. Pure ash is white in colour.

Moisture

A freshly felled tree anything from 40% to 60% of hygroscopic moisture depending upon the species of the tree as well as the seasons of the year. On exposure to atmospheric air, the moisture dries up and reduces to 15-20% in about 18 months. On the exposure for a longer period, no appreciable change had been observed. When wood is seasoned in water, it absorbs nearly 150% of water by weight.

Characteristics of Flame

The nature of the flame depends on the tar content of wood. Pine and birch contain more tar and hence burn with a thick and bright flame, while aspen and alder burn with a dim, transparent flame. The length of the flame also depends on the tar content.

Combustion Characteristics

The lighter the wood, the more intensely it burns with a long flame. This is because air penetrates easily throughout the whole piece during combustion. If the wood is heavy, i.e., hard, the penetration of air is rendered difficult and a concentrated flame results with the development of more heat at the point of burning.

Ignition Temperature

Wood ignites very easily. That is why it is used for lighting other fuels. The average ignition temperature of different kinds of wood is given in Table 7.7.

Table 7.7 Average ignition temperature of wood

Type of wood	Ignition temperature (°C)
Pine	295
Oak	287
Larch	290
Fir	292

7.3.2 Coals and their Characteristics

It is commonly adopted view that coal is a mineral substance of vegetable origin. The large deposits of coal in India are in Bengal, Bihar and Madhya Pradesh. Most of the Indian coal is of low grade variety and coal washing to obtain low ash metallurgical coal is unavoidable. Over 30% of coal output is consumed by railways; another similar proportion is used by industry including iron and steel works. This leaves barely 40% of coal mined for use of the power supply undertakings.

Analysis of Coal

To ascertain the commercial value of coal certain tests regarding its burning properties are performed before it is commercially marketed. Two commonly used tests are : Proximate analysis and Ultimate analysis of coal. Calorific value of coal is defined as the quantity of heat given out by burning one unit weight of coal in a calorimeter.

Proximate Analysis of Coal

This analysis of coal gives good indication about heating and burning properties of coal. The test gives the composition of coal in respect of moisture, volatile matter, ash and fixed carbon. The moisture test is performed by heating 1 gm of coal sample at 104 °C to 110 °C for 1 hour in an oven and finding the loss in weight. The volatile matter is determined by heating 1 gm of coal sample in a covered crucible at 950 °C for 7 minutes and determining loss in weight, from which the moisture content as found from moisture test is deducted. Ash content is found by completely burning the sample of coal in a muffled furnace at 700 °C to 750 °C and weighing the residue. The percentage of fixed carbon is determined by difference when moisture, volatile matter and ash have been accounted for. The results of proximate analysis of most coals indicate the following broad ranges of various constituents by weight:

Moisture	3-30%
Volatile matter	3-50%
Ash	2-30%
Fixed carbon	16-92%

The importance of volatile matter in coal is due to the fact that it largely governs the combustion which in turn governs the design of grate and combustions space used. High volatile matter is desirable in gas making, while low volatile matter for manufacturing of metallurgical coke.

The Ultimate Analysis of Coal

This analysis of coal is more precise way to find the chemical composition of coal with respect to the elements like carbon, hydrogen, oxygen, nitrogen, sulphur and ash. Since the content of carbon and hydrogen that is already combined with oxygen to form carbondioxide and water is of no value for combustion, the chemical analysis of coal alone is not enough to predict the suitability of coal for purpose of heating. However, the chemical composition is very useful in combustion calculations and in finding the composition of fuel gases. For most purposes the proximate analysis of coal is quite sufficient.

The broad ranges in which the constituents of coal vary by weight as determined by ultimate analysis are given below:

Carbon	50-95%
Hydrogen	2.5-5%
Oxygen	2-4%
Sulphur	0.5-7%
Nitrogen	0.5-3%
Ash	2-30%

7.3.3 Manufactured Solid Fuels and their Characteristics

The manufactured solid fuels include, charcoal, coke, briquettes, etc. They are obtained from the natural fuels, like wood, coal, etc.

Charcoal and its Characteristics

Out of the mentioned various manufactured fuels, the charcoal occupies the first place in India. In some parts of the country, for example, Mysore, huge quantities of charcoal are being used till today in blast furnaces for reducing iron ores, etc., and in many homes charcoal is used for cooking purposes. Charcoal is a produce derived from destructive distillation of wood, being left in the shape of solid residue. Charcoal burns rapidly with a clear flame, producing no smoke and developing heat of about 6,050 cal/kg.

Coke and its Characteristics

It is obtained from destructive distillation of coal, being left in the shape of solid residue. Coke can be classified into two categories: soft coke and hard coke. Soft coke is obtained as the solid residue from the destructive distillation of coal in the temperature range of 600-650 °C. It contains 5 to 10% volatile matter. It burns without smoke. It is extensively used as domestic fuel. Hard coke is obtained as solid residue from the destructive distillation of coal in the temperature range of 1200-1400 °C. It burns with smoke and is a useful fuel for metallurgical process.

Briquettes and their Characteristics

The term briquettes is used in respect of the dust, culm, slack and other small size waste remains of lignite, peat, coke, etc., compressed into different shapes of regular form, with or without binder. Dust and rubble result in considerable percentage during mining, transportation, etc., and the briquetting industry is, therefore, an important step towards the saving of fuel economy.

Good briquettes should be quite hard and as little friable as possible. They must withstand the hazards of weather, and must be suitable for storing and general handling in use. These properties are important to briquettes by a correctly selected binder, or suitable processing such as pre-heating, pressing, etc. Amongst the binders, asphalt, pitch are most commonly used, giving fine results. The general conclusion is that 5-8% binder should be used to produce high quality briquettes.

Bagasse and its Characteristics

Bagasse is the residue of sugarcane, left as waste in the sugar mill after extraction of sugar juice. In weight, it is about 20% of virgin cane. By nature, it is fibrous fuel which can be compared to wood. It contains 35-45% fibre, 7-10% sucrose and other combustible, and 45-55% moisture, and possesses an average calorific value of 2200 cal/kg. On moisture-fibre basis the average composition is:

$$C = 45\%, H_2 = 6\%, O_2 = 46\% \text{ and } Ash = 3\%$$

Bagasse is the main fuel satisfying the needs of sugar industries and efforts are being made for decreasing the present moisture of bagasse with the help of flue-gas waste heat dryers. Bagasse is a quick burning fuel with good efficiency.

7.4   Liquid Fuels and their Characteristics

The liquid fuels can be classified as follows:

    (a) Natural or crude oil, and

    (b) Artificial or manufactured oils.

The advantages and disadvantages of liquid fuels can be summarized as follows:

Advantages

    (a) They possess higher calorific value per unit mass than solid fuels.

    (b) They burn without dust, ash, clinkers, etc.

(c)   Their firing is easier and also fire can be extinguished easily by stopping liquid fuel supply.

(d)   They are easy to transport through pipes.

(e)   They can be stored indefinitely without any loss.

(f)   They are clean in use and economic to handle.

(g)   Loss of heat in chimney is very low due to greater cleanliness.

(h)   They require less excess air for complete combustion.

(i)   They require less furnace space for combustion.

Disadvantages

(a)   The cost of liquid fuel is relatively much higher as compared to solid fuel.

(b)   Costly special storage tanks are required for storing liquid fuels.

(c)   There is a greater risk of fire hazards, particularly, in case of highly Inflammable and volatile liquid fuels.

(d)   They give bad odour.

(e)   For efficient burning of liquid fuels, specially constructed burners and spraying apparatus are required.

7.4.1 Petroleum and its Characteristics

Petroleum is a basic natural fuel. It is a dark greenish brown, viscous mineral oil, found deep in earth's crust. It is mainly composed of various hydrocarbons (like straight chain paraffin's, cycloparaffins or napthenes, olefins, and aromatics) together with small amount of organic compounds containing oxygen, nitrogen and sulphur. The average composition of crude petroleum is: C = 79.5 to 87.1%; H = 11.5 to 14.8%; S = 0.1 to 3.5%, N and O = 0.1 to 0.5%.

Petroleum's are graded according to the following phsio-chemical properties:

(a) Specific gravity,

(b) Calorific value,

(c) Flash point or ignition point,

(d) Viscosity,

(e) Sulphur contents,

(f) Moisture and sediment content, and

(g) Specific heat and coefficient of expansion.

Classification of Petroleum

The chemical nature of crude petroleum varies with the part of the world in which it is found. They appear, however, to be three principal varities.

Paraffinic base type crude petroleum

This type of petroleum is mainly composed of the saturated hydrocarbons from CH_4 to $C_{35} H_{72}$ and a little of the napthenes and aromatics. The hydrocarbons from $C_{18} H_{38}$ to $C_{35} H_{72}$ are sometimes called waxes.

Asphaltic base type crude petroleum

It contains mainly cycloparaffins or napthenes with smaller amount of paraffin's and aromatic hydrocarbons.

Mixed base type crude petroleum

It contains both paraffinic and asphaltic hydrocarbons and is generally rich in semi-solid waxes.

7.4.2 Manufactured Liquid Fuels and their Characteristics

Manufactured liquid fuels include Gasoline, Diesel oil, Kerosene, Heavy oil, Naptha, Lubricating oils, etc. These are obtained mostly by fractional distillation of crude petroleum or liquefaction of coal.

Gasoline or Petrol and its Characteristics

The straight run gasoline is obtained either from distillation of crude petroleum or by synthesis. It contains some undesirable unsaturated straight chain hydrocarbons and sulphur compounds. It has boiling range of 40-120 °C. The unsaturated hydrocarbons get oxidized and polymerized, thereby causing gum and sludge formation on storing. On the other hand, sulphur compounds lead to corrosion of internal combustion engine and at the same time they adversely affect tetraethyl lead, which is generally added to gasoline for better ignition properties. The sulphur compounds from gasoline are generally removed by treating it with an alkaline solution sodium plumbite. Olefins and colouring matter of gasoline are usually removed by percolating through 'Fuller's earth' which absorbs preferentially only the colours and olefine. It is used in air-crafts. It is also used as motor fuel, in dry-cleaning and as a solvent.

Some of the characteristics of an ideal gasoline are the following

    (a) It must be cheap and readily available.

    (b) It must burn clean and produce no corrosion, etc., on combustion.

    (c) It should mix readily with air and afford uniform manifold distribution, i.e., should easily vaporize.

    (d) It must be knock resistant.

    (e) It should be pre-ignite easily.

    (f) It must have a high calorific value.

Diesel Fuel and its Characteristics

The diesel fuel or gas oil is obtained between 250-320 °C during the fractional distillation of crude petroleum. This oil generally contains 85% C 12% H. Its calorific value is about 11,000 kcal/kg. The suitability of a diesel fuel is determined by its cetane value. Diesel fuels consist of longer hydrocarbons and have low values of ash, sediment, water and sulphate contents. The main characteristics of a diesel fuel are that it should easily ignite below compression temperature. The hydrocarbon molecules in a diesel fuel should be, as far as possible, the straight-chain ones, with a minimum admixture of aromatic and side-chain hydrocarbon molecules. It is used in diesel engines as heating oil and for cracking to get gasoline.

Kerosene Oil and its Characteristics

Kerosene oil is obtained between 180-250 °C during fractional distillation of crude petroleum. It is used as an illuminate, jet engine fuel, tractor fuel, and for preparing laboratory gas. With the development of jet engine, kerosene has become a material of far greater importance than it is used to be. When kerosene is used in domestic appliances, it is always vaporized before combustion. By using a fair excess of air it burns with a smokeless blue flame.

Heavy Oil and its Characteristics

It is a fraction obtained between 320-400 °C during fractional distillation of crude petroleum. This oil on re-fractionation gives:

    (a) Lubricating oils which are used as lubricants.

    (b) Petroleum-jelly (Vaseline) which is used as lubricants in medicines and in cosmetics.

    (c) Greases which are used as lubricants.

    (d) Paraffin wax which is used in candles, boot polishes, wax paper, tarpolin cloth and for electrical insulation purposes.

7.5 Gaseous Fuels and their Characteristics

Gaseous fuels occur in nature, besides being manufactured from solid and liquid fuels. The advantages and disadvantages of gaseous fuels are given below:

Advantages

Gaseous fuels due to erase and flexibility of their applications, possess the Following advantages over solid or liquid fuels:

(a) They can be conveyed easily through pipelines to the actual place of need, thereby eliminating manual labour in transportation.

(b) They can be lighted at ease.

(c) They have high heat contents and hence help us in having higher temperatures.

(d) They can be pre-heated by the heat of hot waste gases, thereby affecting economy in heat.

(e) Their combustion can readily by controlled for change in demand like oxidizing or reducing atmosphere, length flame, temperature, etc.

(f) They are clean in use.

(g) They do not require any special burner.

(h) They burn without any soot, or smoke and ashes.

(i) They are free from impurities found in solid and liquid fuels.

Disadvantages

(a) Very large storage tanks are needed.

(b) They are highly inflammable, so chances of fire hazards in their use is high.

7.5.1 Natural Gas and its Characteristics

Natural gas is generally associated with petroleum deposits and is obtained from wells dug in the oil-bearing regions. The approximate composition of natural gas is:

$$CH_4 = 70.9\%, \ C_2H_6 = 5.10\%, \ H_2 = 3\%, \ CO + CO_2 = 22\%$$

The calorific value varies from 12,000 to 14,000 kcal/m^3. It is an excellent domestic fuel and is conveyed in pipelines over very large distances. In America, it is available to a great extent, and so, is quite popular as a domestic fuel. It is now used in manufacture of chemicals by synthetic process. It is a colourless gas and is non-poisonous. Its specific gravity is usually between 0.57 to 0.7.

7.5.2 Manufactured Gases and their Characteristics

Manufactured gases are obtained from solid and liquid fuels. Some of the important manufactured gaseous fuels whose characteristics are discussed in the following sections are coal gas, blast furnace gas, water gas, producer gas and oil gas.

Coal Gas its Characteristics

Coal gas is obtained when it is carbonized or heated in absence of air at about 1300 °C in either coke ovens or gas-making retorts. In gas making retort process coal is fed in closed silica retorts, which are then heated to about 1300 °C by burning producer gas and air mixture.

$$C + \frac{1}{2}O_2 \rightarrow CO + 29.5 \text{ kcal}$$

Coal gas is a colourless gas having a characteristic odour. It is lighter than air and burns with a long smoky flame. Its average composition is :

$H_2 = 47\%, CH_4 = 32\%, CO = 7\%, C_2H_2 = 2\%, C_2H_4 = 3\%, N_2 = 4\%, CO_2 = 1\%$ and rest $= 4\%$. Its calorific value is about 4,900 kcal/m^3.

It is used as (a) Illuminate in cities and town, (b) A fuel, and (c) In metallurgical operations for providing reducing atmosphere.

Blast Furnace Gas and its Characteristics

It is a by-product fuel gas obtained during the reduction of iron core by coke in the blast furnace. Its calorific value is about 1,000 kcal/m^3. It contains about 20-25% carbon monoxide along with CO^2, N^2, etc. About 1/3 of this gas is used for preheating air used in blast furnace itself; while the remaining 2/3rd is available for use in boilers or after cleaning in gas engines. It is also used for burning in a special type of stoves (called Cowper's stove) where the furnace is preheated.

This gas contains much dust and is usually cleaned before use by dust settlers, cyclones or electrolytic precipitators.

Water Gas and its Characteristics

Water gas is essentially a mixture of combustible gases CO and H_2 with a little fraction of non-combustible gases. It is made by passing alternatively steam and is little air through a bed of red hot coal or coke maintained at about 900 to 1000 °C in a reactor, which consists of a steel vessel about 3 m wide and 4 m in height. It is lined inside with fire-bricks. It has a cup and cone feeder at the top and an opening at the top for the exit of water gas. At the base, it is provided with inlet pipes for passing air and steam.

Reactions

Supplied steam reacts with red hot coke (or coal) at 900-1000 °C to form CO and H_2.

$$C + H_2O \rightarrow CO + H_2 - 29 \text{ kcal}$$

$$C + O_2 \rightarrow CO_2 + 97 \text{ kcal}$$

Composition

The average composition of water gas is: H_2 = 51%; CO = 41%; N_2 = 4%; CO_2 = 4%. Its calorific value is about 2,800 kcal/m^3.

Uses

It is used as (a) a source of hydrogen gas, (b) an illuminating gas, and (c) a fuel gas.

Producer Gas and its Characteristics

Producer gas is essentially a mixture of combustible gases carbon monoxide and hydrogen associated with non-combustible gases N_2, CO_2, etc. It is prepared by passing air mixed with little steam (about 0.35 kg/kg of coal) over a red hot coal or coke bed maintained at about 1100 °C in a special reactor called gas producer. It consists of a steel vessel about 3 m in diameter and 4 m in height. The vessel is lined inside with fire bricks. It is provided with a cup and cone feeder at the top and a side opening for the exit of producer gas. At the base it has an inlet for passing air and steam. The producer at the base is also provided with an exit for the ash formed.

Reactions

The gas production reactions can be divided into four zones as follows :

Ash Zone

The lowest zone consists of mainly of ash, and therefore, it is known as ash zone.

Combustion Zone

The zone next to the ash zone is known as oxidation or combustion zone. Here the carbon burns and forms CO and CO_2. The temperature of this zone is about 1100 °C. The following reactions take place.

$$C + O_2 \rightarrow CO_2 + 94 \text{ kcal}$$

$$C + \frac{1}{2}O_2 \rightarrow CO + 29.5 \text{ kcal}$$

Reduction Zone

Here carbon dioxide and steam combines with red hot carbon and liberates free hydrogen and carbon monoxide. The reactions are:

$$CO_2 + C \rightarrow 2CO - 94 \text{ kcal}$$

$$C + H_2O \rightarrow CO + H_2 \rightarrow 29 \text{ kcal}$$

$$C + 2H_2O \rightarrow CO_2 + 2H_2 - 19 \text{ kcal}$$

All these reduction reactions are endothermic, so, the temperature in the reduction zone falls to 1000 °C.

Distillation Zone

In this zone (400 – 800 °C) the incoming coal is heated by outgoing gases by giving sensible heat to the coal. The heat given by the gases and heat radiated from the reduction zone helps to distillate the fuel thereby volatile matter of coal is added to the outgoing gas.

Composition

The average composition of producer gas is CO = 22.3%, H_2 = 8.12%; N_2 = 52.55%; CO_2 = 3%. Its calorific value is about 1,300 kcal/m^3.

Uses

It is cheap, clean and easily preparable gas and is used (i) for heating open-heat furnaces (in steel and glass manufacture), muffle furnaces, retorts (used in coke and coal gas manufacture), etc., and (iii) as a reducing agent in metallurgical operations.

Oil Gas and its Characteristics

Oil gas is obtained by cracking kerosene oil. Oil in a thin steam is allowed to fall on a stout red hot cast iron retort, which is heated in coal fired furnace. The resulting gaseous mixture passes out through a bonet cover to a hydraulic main, a tank containing water. Here tar gets condensed. Then at the testing cap, the proper cracking of oil is estimated from the colour of the gas produced. A good oil gas should have a golden colour. By proper adjusting the supply of air, gas of required colour can be obtained. The gas is finally stored over water in gas holders.

Composition

The average composition of oil gas is : CH_4 = 25.30%; H_2 = 50-55%;

CO = 10.12%; CO_2 = 3%. Its calorific value is about 6,600 kcal/m^3.

Uses

It is used as laboratory gas.

7.6   Liquefied Petroleum Gas (LPG)

It is lighter type of hydrocarbon with molecular formula C_3H_6. It is a by-product of petroleum refining or natural gas processing.

Properties

1. LPG is a dry gas under atmospheric conditions, but can be liquefied under pressure.
2. LPG consists of mainly propane (80%) and some proportion of butane (20%)
3. It has Octane number about (103-105)
4. Its heating value is about 48 MJ/kg.

Advantages

1. Complete air fuel mixing at all temperature.
2. Complete vaporization
3. Closer control of air-fuel ratios.
4. Uniform mixture to all cylinders in a multi-cylinders engine.
5. No condensed liquid will dilute the lubrication oil because of vapor form. This means lesser oil consumption.
6. No carbon or Gum deposit formed on engine parts
7. Combustion is clean and complete.
8. High anti knock value and doesn't pre-ignite easily.
9. It gives better manifold distributions, which means lesser emissions.
10. Increased engine life.
11. Less engine wears.
12. Lower maintenance
13. It is cheaper than gasoline.
14. Higher durability of exhaust systems and fuel enrichment is automatically eliminated
15. Engines having high compression ratio (10:1) can use LPG with excellent results
16. Manifold heat can be eliminated as fuel is completely vaporized before it reaches the carburetor and will not condense on the way to the engine.

Disadvantages

1. Storage on board and require more space.

2. Handling and storing in cylinder under pressure of about 18 bar.

3. Heat energy is less than that of gasoline (about 80%)

4. It reduces volumetric efficiency due to its high latent heat of vaporization.

5. Its characteristic odor is faint. So leakage cannot be easily detected.

6. Response to blending is poor.

7. Special fuel system has to be provided.

8. Heavy pressure always increases the vehicle weight unnecessarily.

9. Average temperature of gases inside the cylinder is more and this could cause increased thermal loading.

10. Harder to start in winter.

LPG in SI engine (LPG fuel feed system)

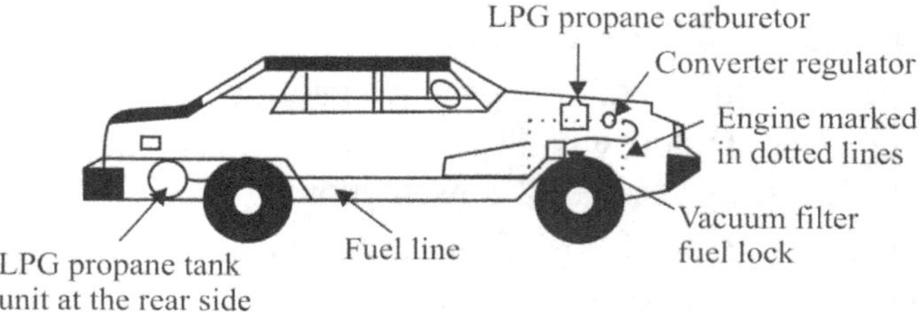

Fig. 7.1 Typical LPG Propane Fuel Feed System of an Automobile Vehicle

In some current designs the fuel tank pressure is used to force the fuel through a filter and in some cases through a solenoid lock off valve actuated; by the ignition switch before reaching the first or primary pressure regulator. Here, the pressure reduction, from 100 psi to about 10 psi causes the liquid to vaporize. LPG vapors move to the LPG propane carburetor. This carburetor supplies the A.F mixture to the engine for combustion (Fig. 7.1).

In some installations heat is supplied by an engine cooling system heat exchange unit to prevent a temperature drop at the point of vaporization. The dry gas then passes to a low-pressure regulator where it is released at a pressure just below atmospheric pressure. These controls insure a supply of fuel to the engine only when it is operation. A special carburetor meters the low-pressure fuel and mixes it with air in the proper proportions.

LPG in CI Engines

LPG can also be used in CI engine by introducing the gas into inlet manifold. Since this fuel has poor auto ignition quality. It can only be used as an auxiliary fuel for dual diesel engines. Diesel oil will be injected in the conventional manner and with increasing load on the engine. The quantity of diesel oil is gradually reduced by the action of governor. The inducted LPG will turn by flame propagation originating from the combustion process will depend on strength of the LPG air mixture and the distribution.

Engine Modifications

By means of cooling and compression they are condensed to a liquid and stored in a sealed pressure tanks.

Their extreme volatility under reduced pressure requires the use of special fuel system equipment to control their flow from fuel tank to the engine cylinders, as they are completely vaporized.

Equipment for Refilling the Tank is necessary

LPG auto conversion kits are available which contain axillaries like the converter or pressure regulator.

An ignition setting has to be changed. Since LPG is dry fuel has higher electrical resistance, spark plug voltage has to be increased 30-40%. An outcome in these spark plugs has to be changed more frequently.

A highly desirable addition to fuel system is a shut down valve actuated by the ignition key so that the gas flow is stopped automatically when engine is switched off.

Application

1. Its largest use is at present as domestic and industrial fuel.
2. It is more widely used by bus lines, cab fleets, tractors, truck fleets, railroad, and industrial operation.
3. It is also used in dual-fuel diesel engines.

Table 7.8 Comparison of Petrol and LPG Fuel

Petrol	Liquefied Petroleum Gas (LPG)
Fuel consumption in petrol is less when compared to LPG	Compared to petrol, running the engine on LPG results in around a 10% increase in consumption
Petrol has odor	LPG is odorless
Octane rating of petrol is 81	Octane rating of LPG is 110
Petrol engine is not as smooth as LPG engine	Due to higher octane rating the combustion of LPG is smoother and knocking is eliminated and the engine runs smoothly.
In order to increase octane number petrol requires lead additives	LPG is lead free with high octane number
The life of petrol engine is shorter	The life of the engine is increased to 50%
Due to formation of carbon deposits on the spark plugs, the life of spark plus is shortened.	Due to absence of carbon deposits on the electrodes of plugs, the life of the spark plug is increased.
Carburetor supplies the mixture of petrol and air in the proper ratio to the engine cylinders for combustion	The vaporizer functions as the carburetor when the engine runs on LPG.

7.7   Natural Gas as Alternate Fuel

Natural gas is a mixture of components, consisting mainly of methane (60-95%) with small amounts of other hydrocarbon fuel components. But usually contain a large amount of methane (CH_4) and small amount of ethane C_2H_5. In addition it contains various amounts of N_2, CO_2 and traces of other gases. Its sulphur content ranges from very little (sweet) to larger amounts (sour).

Natural gas obtained from oil wells is called casing head gas. It is usually treated for the recovery of gasoline. After this, it is called dry gas. It is delivered into the pipeline systems to be used as fuel. Natural gas can be used in the production of natural gasoline.

It is stored as compressed natural gas (CNG) at pressures of 16 to 25 bar, or as liquid natural gas LNG at pressures of 70 to 210 bar and a temperature around −160 °C. As a fuel, it works best in an engine system with a single-throttle body fuel injector. This gives a longer mixing time, which is needed by this fuel.

Advantages of Natural Gas

   (i) Octane number is around 120, which makes it a very good SI engine fuel.

  (ii) Low engine emissions. Less aldehyde than with methanol.

 (iii) Fuel is fairly abundant worldwide

 (iv) High flame speed engines can operate with a high compression ratio.

Disadvantages of Natural Gas

   (i) Low energy density resulting in low engine performance.

  (ii) Low engine volumetric efficiency because it is a gaseous fuel.

 (iii) Need for large pressurized fuel storage tank. There is some safety concern with a pressurized fuel tank.

 (iv) Inconsistent fuel properties.

  (v) Refueling is a slow process.

Compressed Natural Gas CNG

CNG is used to run an automobile vehicle just like LPG. The CNG fuel feed system is similar to the LPG fuel feed system. CNG conversion kits are used to convert petrol driven cars into CNG - driven cars. These kits contain auxiliary parts like the converter, mixer and other essential parts required for conversion in an engine system with a single-throttle body fuel injector.

Natural gas in CI engines

Some very large stationary CI engines operate on a fuel combination of methane and diesel fuel. Methane is the major fuel, amounting to more than 90% of the total. It is supplied to the engine as a gas through high-pressure pipes. A small amount of high grade, low sulfur diesel fuel is used for ignition purposes. The net result is very clean running engines.

7.8 Hydrogen as Alternate Fuel in IC Engine

Properties

Boiling point: $-253\,^\circ$C, Heat of vaporization: 443.2 kJ/kg at $20\,^\circ$C Lower heating value: 119.9 MJ/Kg; ignition temperature: $585\,^\circ$C; stoichiometric mass ratio 34:3.

1. It is the simplest and most common of elements.

2. It is in gases state at atmospheric pressure and ambient temperature.

3. Hydrogen contain low level of carbon monoxide and carbon dioxide depending upon the source.

4. Hydrogen has very low energy density.

Advantages

1. It can be manufactured from unlimited source.

2. Clean burning properties as it produces only water vapour and small amount of NO_x.

3. Thermal efficiency is higher because of higher compression ratio.

4. Lean mixture can be burned more efficiently because of wider flammability limits.

5. Hydrogen diffuses rapidly to produce a uniform mixture in the combustion chamber.

6. Since there is no condensation of fuel on cold metal parts, short range cold driving is possible.

7. Reduced green house emissions.

8. Stronger national energy security.

Hydrogen can be used in SI engines by three methods

1. By manifold induction.

2. By direct introduction of hydrogen into the cylinder.

3. By supplementary to gasoline.

In the manifold introduction of hydrogen cold hydrogen is introduced through a valve controlled passage into the manifold. This helps to reduce the risk of back flash. The power output at the engine is limited by two factors - preignition and back flash. Also the energy content of air hydrogen mixture is lower than that of liquid hydrocarbon fuels.

In the direct introduction of hydrogen into the cylinder hydrogen is stored in the liquid forms, in a cryogenic cylinder. A pump, pumps this liquid through a small heat exchanger where it is converted into cold hydrogen gas. The metering of the hydrogen is also done in this unit. The cold hydrogen helps to prevent pre-ignition and also reduces NO_x formation.

Hydrogen can also be used as a supplementary to gasoline in SI engines. In this system, hydrogen is inducted along with gasoline compressed and ignited by a spark.

The arrangements of liquid hydrogen storage and details of hydrogen introduction into the SI engine cylinder can be seen in Fig. 7.2.

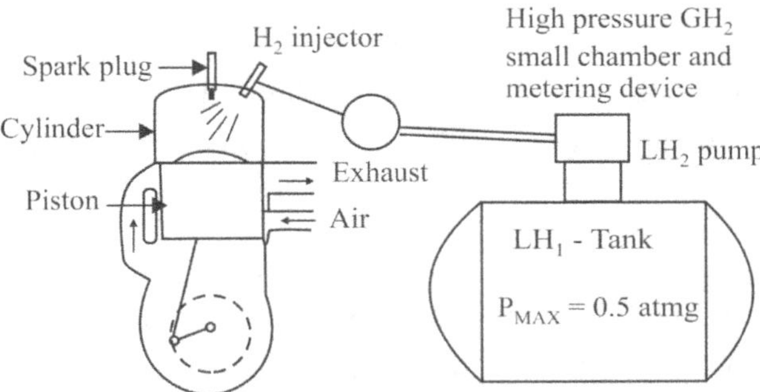

Fig. 7.2 Hydrogen SI Engine System

There are two methods by which hydrogen can be used in diesel engines

1. By introducing hydrogen with air and using a spray of diesel oil to ignite the mixture that is by the dual fuel mode. The limiting conditions are when the diesel quantity is too small to produce effective ignition, that is failure of ignition and when the hydrogen air mixture is so rich that the combustion becomes unacceptably violent. In between these limits a wide range of diesel to hydrogen proportion can be tolerated. Investigations show that beyond a certain range (30% to 50% substitution of diesel fuel by hydrogen), further introduction leads to violent pressure rise.

2. By introducing hydrogen directly into the cylinder at the end of compression. Since the self-ignition temperature of hydrogen is very high, the gas spray is made to impinge on a hot glow plug in the combustion chamber – that is by surface ignition. It is also possible to feed a very lean hydrogen air mixture during the intake into an engine and then inject the bulk of the hydrogen towards the end of the compression stroke.

Since hydrogen is a highly reactive fuel it requires great care in handling. Flash black arresters have to be provided between the engine and the storage tank to prevent flash back from going to the tank.

7.9   Alcohols as a Alternate Fuels for IC Engines

Alcohols are an attractive alternate fuel because they can be obtained from both natural and manufactured sources.

Advantages

(i)   It can be obtained from a number of sources, both natural and manufactured.

(ii) It is high-octane fuel with anti-knock index numbers over 100.

(iii) It produces less overall emissions when compared with gasoline.

(iv) When alcohols are burned, it forms more moles of exhaust gases, which gives higher pressure and more power in the expansion stroke

(v)   It has high latent heat of vaporizations, which results in a cooler intake process. This raises the volumetric efficiency.

(vi) Alcohols have low sulphur content in the fuel.

(vii) High self-ignition temperature.

Disadvantages

(i)   Lower heating values leading to a larger fuel tank and also increased jet sizes in the carburetor.

(ii) Combustion of alcohols produces more aldehydes in the exhaust.

(iii) Alcohol is much more corrosive than gasoline.

(iv) It has poor cold weather starting characteristics due to low vapour pressure and evaporation.

(v)   Alcohols have poor ignition characteristics in general.

(vi) Alcohols have almost invisible flame, which is considered dangerous when handling fuel.

(vii) Because of low flame temperatures, catalytic converter is not heated quickly to an efficient operating temperature.

(viii) There is danger of storage tank flammability due to low vapor pressure.

(ix) Strong odor of alcohol causes headache and dizziness when refueling.

(x)   There is a possibility of vapor lock in the fuel delivery system.

Spark - ignition engine modifications generally require to utilize alcohols (methanol and Ethanol fuels with good fuel) economy and operation are:

1.   Carburetor re-calibration to produce and maintain the required fuel air ratios.

2.   Materials in fuel systems may need to be changed.

3. Increases heat on intake manifold to provide sufficient capacity (3.7 to 6.4 times higher) to vaporize the alcohol to retain good cylinder to cylinder distribution and over all operation.

4. Modify ignition advance. Usually the optimum timing will be different due to different combustion characteristics of gasoline and alcohols.

5. Providing starting aids either an auxiliary gasoline starting system or a means of heating the intake manifold. Un-aided alcohol fuels, with their high latent heat of vaporizations will not vaporize sufficiently enough to start the engine under all conditions.

6. An increase in compression ratio will provide improved fuel economy of engine. But this is not mandatory for operation. It is desirable to increase the ratio to take advantage of the higher octane number of the alcohols.

7. In some cases, Lubrication of valves and upper piston ring area may need to be modified to prevent excess wear due to alcohols fuels acting as solvent and washing the lubrication film from the critical areas.

8. The fuel tank capacity will have to be increased by 1.5 to 1.9 times depending on ratio of ethanol. The ratios will be higher for methanol.

9. Corrosion resistant materials are required for fuel system as alcohols are corrosive in nature.

Some engines have been converted to use alcohol fuels by only modifying the carburetor. However, the operation is generally not accepted due to starting difficulty. To obtain good operation, nearly all of the above changes are required.

The conversion of engine to operate straight on alcohol fuels means that it can no longer operate on gasoline without reconverting it. Basically, there is no differences between methanol and ethanol, so for their suitability for engines is considered.

Blends with Gasoline: The possibility of mixing methanol or ethanol with gasoline has received considerable attention. The primary advantages are the larger octane boosting properties of methanol and ethanol and potential of diminishing gasoline supplies. High blending octane ratings of methanol and ethanol would make effective replacement of conventional lead additives. The blends give improved efficiency and reduced emissions.

Alcohols for CI Engines

Techniques of using alcohols in diesel engines are:
1. Alcohol/diesel fuel solutions.
2. Alcohol diesel emulsions.
3. Alcohol fumigation.
4. Dual fuel injection.
5. Surface ignition of alcohols.
6. Spark ignition of alcohols.
7. Alcohols containing ignition improve additives.

- Ignition improves additives for increasing the cetane number are available such as Amyl Nitrate, Hexyl Nitrate, Cyclohexyl Nitrate etc. In diesel fuels a few tenth of one % constitute a treatment. The engine operates with greater thermal efficiency using the alcohol/cetane improver blend.

- In the dual fuel engine, alcohol is carbureted or injected into the inducted air. Due to high self-ignition temperature of alcohols there will be no combustion with usual diesel compression ratios of 16 to 18. A little before the compression stroke, small quantity of diesel oil is injected into combustion chamber through the normal diesel pump and spray nozzle. The diesel oil readily ignites and this initiates combustion in the alcohol air mixture. Several methods are adopted for induction of alcohol into the intake manifold. They are micro fog unit, pneumatic spray nozzle, vaporizer, carburetor and fuel injector.

- *Surface ignition engines:* A slab of insulator material, wound with a few strands of heating wire is fixed on the combustion chamber with the fire running on the face exposed to the gases. The Fuel injector is located such that a part of the spray impinges on this surface. Ignition is thus initiated. The combustion chamber, which is in the cylinder head, is made relatively narrow so that the combustion spreads quickly to the rest of the space. Since a part of the fuel burns on the insulator surface and since the heat losses from the plate are low, the surface after some minutes of operation reaches a temperature sufficient to initiate ignition without the aid of external electrical power supply. The power consumption of the coil is about 50 to 60 watt. The engine lends itself easily to the use of wide variety of fuels, including methanol, ethanol and gasoline. The engine operates more smoothly at lower speeds than at higher speed (Fig.7.3).

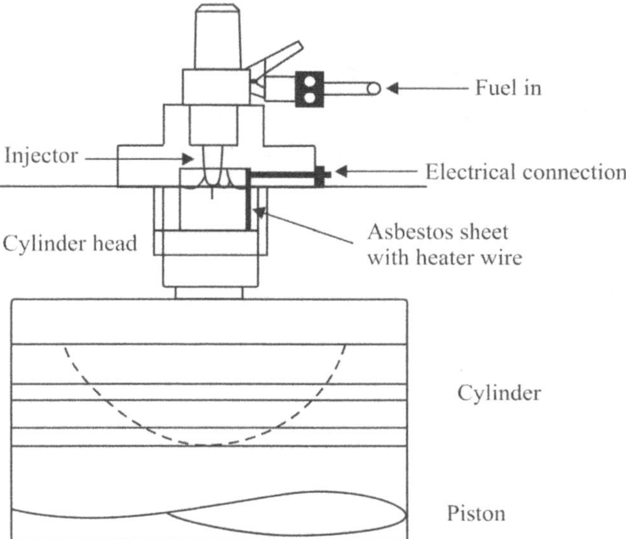

Fig. 7.3 Alcohol Surface-ignition Engine

- *Spark assisted diesel:* A Spark assisted diesel engine can operate with future low cetane fuel without losing the characteristics of diesel engines. A conventional pre-combustion chamber type diesel engine can be modified by installing spark plug in its pre-combustion chamber. Road performance of such type vehicle is found to be good. These vehicles have a cleaner exhaust with smoke and odour eliminated. Oxides of nitrogen and particulates are also reduced.

Review Questions

1. Define LPG. What are the Properties of LPG which are in favour to use as an LPG Engine fuel? What Modification are required to use as a substitute fuel? Explain LPG fuel feed system. Compare LPG and Petrol as fuel for SI Engine?

2. What is natural gas? What are the advantages and disadvantages of using natural gas as alternate fuels?

3. Explain the two methods by which H_2 can be used in SI and CI Engine. What are the advantages and disadvantages of using hydrogen?

4. Explain alcohols as a alternate fuels for IC engines bringing out their merits and demerits. Compare the properties of alcohols and gasoline as engine fuels. What modifications are required with engine system?

CHAPTER 8

Recent Trends in
IC Engines

- Lean burn Engines
- Stratified Charge Engines
- Gasoline Direct Injection Engine
- Plasma Ignition
- Homogeneous charge compression Ignition
- Four Valve and overhead Cam Engine

8.1 Homogenous Charge Compression Ignition (HCCI) Engine

Homogeneous charge compression ignition or HCCI is a relatively new combustion technology. HCCI is a hybrid of traditional diesel (direct injected compression ignition) engines and gasoline (conventional spark ignition) engines. Unlike a traditional SI or Diesel engine, HCCI combustion takes place spontaneously and homogeneously in air/fuel mixture regions. In addition, HCCI is a lean combustion process. HCCI is an ideal combination of an engine type with efficiency of CI engine and very low emissions of that of SI engine.

Working Principle of HCCI

In a Homogeneous Charge Compression Ignition (HCCI) engine, the fuel is premixed into the incoming air to create a homogeneous charge before the combustion. HCCI engines operate on the principle of having a dilute, premixed charge that reacts and burns volumetrically throughout the cylinder as it is compressed by the piston. This early injection allows the fuel to homogeneously mix with the air as it is being compressed. However, instead of relying on a spark to ignite the mixture, HCCI relies on the heat of compression to ignite the mixture (Fig. 8.1). The density and temperature of the mixture are raised by compression until the entire mixture reacts spontaneously.

In the SI engine there are three zones, a burnt zone, an unburned zone and between them a thin reaction zone where the chemistry takes place. This reaction zone propagates through the combustion chamber and thus we have flame propagation. Even though the reactions are fast in the reaction zone, the combustion process will take some time as the zone must propagate from spark plug to the far liner wall. With the HCCI process the entire mass in the cylinder will react at once. HCCI has no flame propagation instead the whole mixture burns close to homogeneous during combustion

During the compression stroke, the charge is heated to obtain auto-ignition, preferably close to TDC. The mixture burns simultaneously throughout without the hot flame front of spark ignition combustion or the locally rich flame front of compression ignition combustion. The process also allows for a uniform "auto ignition," or combustion without the need of a spark, at a lower compression than normally required for diesel engines.

In SI engines large cycle-to cycle variations occur since the early flame development varies significantly. The cycle-to cycle variations of the HCCI combustion process are very small since combustion initiation occurs in many places simultaneously.

Features of HCCI

(a) The temperature after compression stroke should equal the auto ignition temperature of the fuel/air mixture. The mixture should be diluted enough to give reasonable burn.

(b) Fuel distribution is likely to be an important parameter for the combustion in an HCCI engine.

(c) Ignition process relies on spontaneous auto-ignition.

(d) As the whole bulk burns almost simultaneously the combustion rate becomes very high. Therefore highly diluted mixtures have to be used in order to limit the rate of combustion.

(e) By adjusting the remote operating parameters e.g., Gas temperature, compression ratio, air/fuel ratio or exhaust gas recirculation (EGR) rate, combustion process can be controlled.

(f) Exhaust gas recirculation can also be used to control and improve the performance of this engine.

Direct injection Diesel engine with a fixed compression ratio can be converted to run in the HCCI Mode. The intake air can be preheated with an air heater. As of today, various fuels such as propane, butane and methane have been tested on this engine with some positive results. Next to HCCI, Controlled Auto-Ignition (CAI) is a name that is used frequently. Furthermore, the same name might be used for different processes.

Advantages of HCCI

Fuel-flexibility: can be operated with wide range of fuels.

Low NO_x and Particulate matter emissions .

Fewer problems with smoke depending on the fuel used.

Easier way to reduce both NO_x and soot simultaneously through combustion improvement.

Higher efficiency at low and part load.

Reduced fuel consumption.

Efficiency can be improved by 15 percent to 20 percent of gasoline engines.

Reduced engine wear and tear.

Disadvantages

The major problem is controlling the ignition timing over a wide load and speed range and to obtain an acceptable power density.

More difficult to control combustion process.

Power density is limited by combustion noise and high peak pressures.

High in-cylinder peak pressures and heat release may cause damage to the engine.

Higher emissions of Carbon monoxide & unburned hydrocarbons.

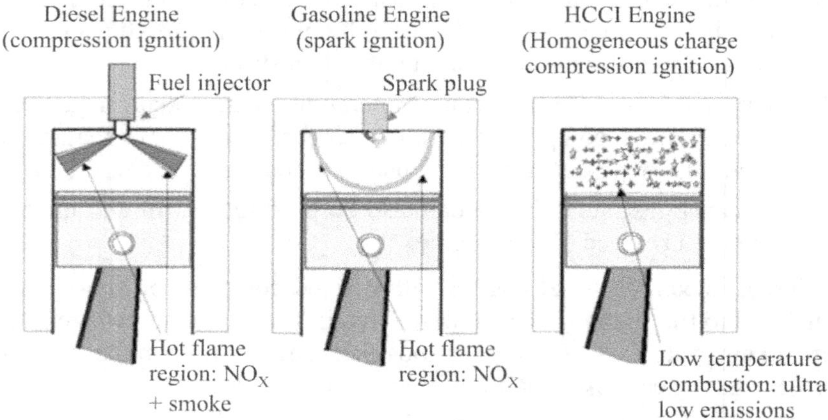

Fig. 8.1 Homogenous Charge Compression Ignition

8.2   Plasma-Jet Ignition System

A good ignition system requires an efficient ignition of varying air fuel ratio supplied to the engine under different conditions of load and speed with no misfire.

In the plasma jet ignition, the spark discharge is confined within a cavity surrounding the plug electrodes. The cavity is connected with the combustion chamber via an orifice the electrical energy supplied to the plug electrodes are much more than the values in the normal ignition voltage and high current. The high power spark discharge creates high temperatures plasma (10,000 to 30,000) so that pressure in the cavity increases substantially which causes a portion of it to be ejected out of the end of the cavity. From where it flows to the main combustion chamber as a turbulent jet, the ignition starts and a turbulent flame proceeds. The penetration of the jet depends on amount of electrical energy, cavity size, cavity volume, orifice area, ambient gas pressure and quantity of fuel present in the cavity. In this system, the flame development period is considerably shortened, ability to burn lean mixture increase and there is no misfire.

Plasma jet is a method for initiating combustion in SI engine with electronic discharge which includes different design of spark plug, use of higher power, high energy, or longer duration spark discharges which initiate the main combustion process with a high temperature reacting jet. The air fuel mix inside the combustion chamber is ignited when an ionized plasma channel traces a path across the spark plug gap. The spark is the 'hottest' as it breaks across the air gap to the other electrode and the fuel is ignited only during this initial break down phase which lasts less than 100 microseconds. The Schematic of plasma jet spark plug is shown in Fig. 8.2.

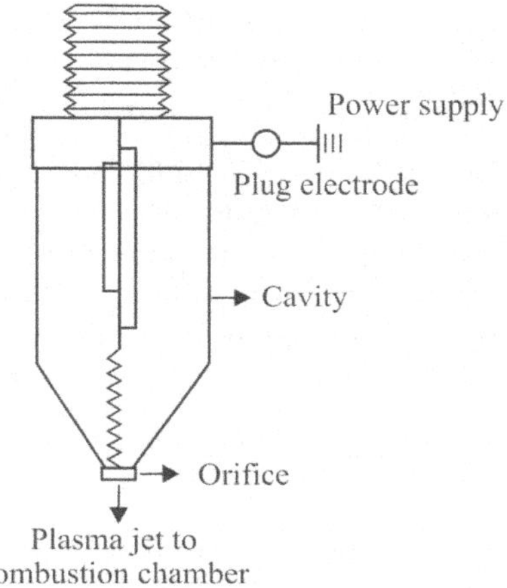

Fig. 8.2 Plasma Jet Ignition

As soon as the spark reaches the ground electrode the subsequent tail end of the spark is of little to no use in the combustion process and is wasted energy. Once the fuel ignites the efficiency or net energy content of the burn is determined by how fast the flame front expands and how much of the fuel is fully combusted. Full combustion results in a greater amount of heat energy liberated which is then converted to mechanical work by the piston.

The rate at which flame front expands is determined by the amount of energy delivered in the first few microseconds as the gap breaks down. A stronger energetic pulse results in a much larger faster flame front while a weak spark results in a flame front that is lower and in some cases only partially ignited.

8.3 Lean Burn Engine

Lean burn engine is a layout of Otto cycle engine designed to permit the combustion of lean air fuel mixtures and to obtain simultaneously low emission values as well as high fuel economy. Generally, all SI engines operate with mixtures nearby stoichiometric air fuel ratio about 12:1 to 16:1. The lean burn engine is designed to operate effectively in the air fuel ratio 14:1 - 16:l to 20:1 - 22:1.

The combustion is considered "lean" when excess air is introduced into the engine along with the fuel. This produces two positive effects. First, the excess air reduces the temperature of the combustion process and this reduces the amount of oxides of nitrogen (NO_x) produced by nearly half. Second, since there is also excess oxygen available, the combustion process is more efficient and more power is produced from the same amount of fuel.

Relation of such an engine has been hampered by the ignition system and combustion aspects, which deteriorate with increasing air fuel ratio. Further mixture misdistribution affects power output between cylinders. When parameters like compression ratio, combustion chamber shape, ignition system, are successfully optimized, the engine is referred to as a lean burn engine.

Working Principle of Lean Burn Engine with Electronic Control System

A lean-burn engine control system uses in-cylinder sensors to monitor combustion pressure, and a sequential injection system to control fuelling on a cylinder-by-cylinder basis. The A/F ratio for the engine is monitored through a wide-range oxygen sensor installed in the exhaust down-pipe. The lean burn engine with electronic control system can be seen in Fig. 8.3.

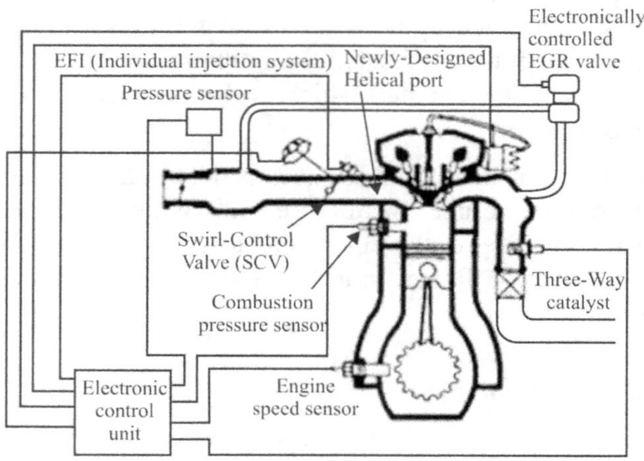

Fig. 8.3 Lean-Burn Engine with ECU

A swirl control valve (SCV) mounted in the intake system. The intake tract for each cylinder is divided into two passages. One passage is smooth, which allows maximum gas flow for good cylinder charging, and the other passage is fitted with a corkscrew-shaped flange, which introduces swirl into the incoming air. The engine electronic control unit (ECU) can switch air flow in one of the two tracts by actuating the SCV. The engine operates in a lean burn mode when it is running under light to medium load conditions. The ECU directs the SCV to open-up the curved inlet passage to have well mixed incoming vapour due to a high level of turbulence through swirl. During combustion, the ECU monitors the pressure sensor signals for any signs of slow or incomplete burning and then appropriately modifies the A/F ratio or ignition timing on a cylinder-by-cylinder basis. Because of the continuous monitoring of the exhaust gas oxygen concentration, these facilities permit the engine to produce high efficiency at A/F ratios upto 25:1. When the engine runs under high load conditions, specifically during overtaking or when hill climbing, the ECU switches to stoichiometric operation for maximum power. To achieve, this ECU directs the SCV to open the smooth intake passage and uses the oxygen sensor signal to maintain an A/F ratio of 14.7:1.

Advantages of lean burn engine

1. *Lower pollutants:* Investigations show that in the lean mixture range all the three major pollutant (CO, HC and NO_x) show descending trend. The drop in temperature after ignition, lowers both NO_x and CO, HC can also be decreased by means of better design parameters like combustion chamber, ignition system and mixture distribution.

2. *Good fuel economy:* With lean burn engines, using lean mixtures, the output torque is moderate till a particular air fuel ratio after that heavy power loss causes adverse effects on the vehicle and also increases in the fuel consumption take place. So with lean burn engine SFC is lowest.

3. *The ratio of specific heat:* (k) approaches that of air with lean mixtures. Dissociation losses are reduced because of lower peak combustion temperatures.

4. Heat transfer losses to the cooling medium are reduced because of lower peak temperatures.

5. Since lean mixtures are less prone to knocking, higher compression ratio can be used which in turn will improves the thermal efficiency.

The following are some of the engine modifications to be made to convert an existing engine as a lean burn engine.

(i)   Increasing the compression ratio of the engine to accelerate flame propagation. Since lean mixtures are resistant to knocking and propagation, this measure can be used.

(ii)   Increasing the swirl and turbulence of the mixture in order to increase the flame speed. This is brought about by inlet port design, and by suitable combustion chamber design

(iii)   Minimizing the heat losses from the combustion chamber and locating the combustion chamber and spark plug near the hot regions of the engine in order to bring about some mild preheating of the mixture.

(iv)   By using an ignition system with high sparking energy and prolonged spark duration.

(v)   Catalytic converter of the charge in the combustion chamber

8.4   Stratified Charge Engines

The two internal combustion engines are petrol engine and diesel engines; the petrol engine has very good full load power characteristics. i.e., high degree of air utilization, high speed and flexibility but rather poor part load efficiency whereas the diesel engine has good part load characteristics but has poor air utilizations, low stroke limited power and higher weight to power ratio. Therefore there is need to develop an engine which combines the advantages of both petrol and diesel engines and at same time avoids as far as possible their disadvantages. Stratified charge engine is an attempt in this direction. It is an engine which is midway between the homogenous charge spark ignition engine and the heterogeneous charge compression ignition engine. The schematic of IC engine in stratified charge operation (air/fuel ratio>14.7:1) and homogenous operation (air/fuel ratio =14.7:1) is shown in Fig.8.4.

(a) Stratified mode   (b) Homogeneous mode

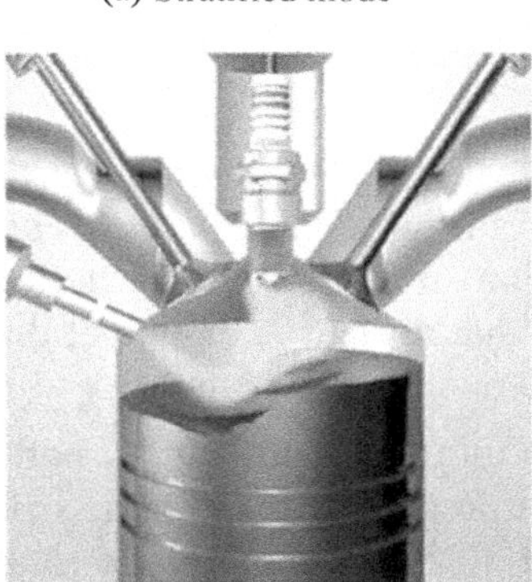
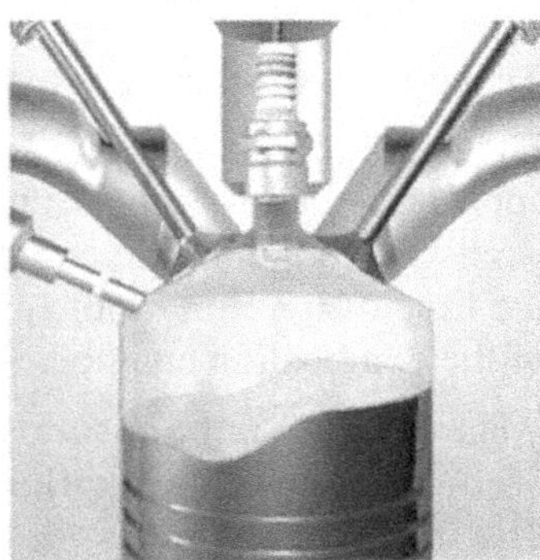

Fig. 8.4 IC Engine in Two Modes of Operation

Charge stratification means providing different fuel air mixture strength at various places in the combustion chamber a relatively rich mixture at and in the vicinity of the spark and a leaner mixture in the rest of the combustion chamber. The stratified charge is usually defined as a spark ignition internal combustion engine in which the mixture in the zone of spark plug is very much richer that in the rest of the combustion chamber. Such engine neither requires high octane number fuel as in SI engines, not require fuel having high ignition quality as in CI engine, and can be operated with a wide range of liquefied fuels.

Characteristics of Stratified Charge Engines

1. All type of stratified charge engine has good part load efficiency while the full load performance is either small or equivalent to petrol engine.
2. Inherent degree of knock resistance, smooth combustion multifuel capacity.
3. Relatively high compression ratio.
4. Ability of direct cylinder fuel injection.
5. Volumetric efficiency of the engine is higher.
6. Exhaust emission characteristics are good.
7. Overall lean air fuel ratio than normal carbureted engine.

Methods of Charge Stratification

The stratified charge engines can be classified into two types.

(a) Those using fuel injection – ford combustion process (FCP) Texaco combustion Process (TCP)

(b)   Those using carburetion – Honda compound vortex controlled combustion (CVCC)

(a) Texaco Combustion System

The air gets a circular swirl during the suction stroke by appropriate port designs and by shrouded inlet valve. The swirl persists throughout the compression stroke and can be increased at the time of ignition by the piston that decreases the upper diameter of the combustion chamber (conversion of momentum).

The nozzle starts injecting the fuel, at about 30 °C before top dead centre into the swirling air (downstream and across stream). The injection continues for about 30 °C of crank angle rotation at full load (smaller quantities of fuel and shorter injection duration at part load). The spark plug is located as close to the nozzle as possible in the downstream position and spark discharge occurs immediately after the injection starts. The flame propagates forming a burning zone downstream from the nozzle. The burning and after burning zone is small at part loads, increases in size with load, and it fills the combustion chamber at full load. The burning zone is relatively rich in fuel and therefore, the flame propagation is similar to the petrol engine. The engine can be supercharged to increase the output.

(b) Honda Compound Vortex Controlled Combustion

This is an example of stratification obtained by carburetion alone. The basic structure of the CVCC engine is the same as that of a conventional 4-stroke gasoline engine except for an auxiliary combustion chamber around the spark plug and the small additional intake valve, which is fitted to each cylinder. The combustion sequence is as follows:

1.   A large amount of lean mixture and a small amount of rich mixture makes the overall lean, is supplied through main and auxiliary intake valves, respectively.

2.   At the end of compression stroke, there is a rich mixture around the spark plug. A moderate mixture in the vicinity of outlet of the auxiliary chamber and a lean mixture in the rest of the main combustion chamber.

3.   The rich mixture in the auxiliary combustion is assured to be ignited by the spark plug.

4. The burning of A/F mixture around the outlet of the auxiliary combustion chamber ensures the combustion of the lean mixture in the main chamber.
5. Lean mixture continues to burn slowly.
6. The temperature of burned gas is kept relatively high for a long duration initiation occurs in many places simultaneously.

Merits

1. Low exhaust emission.
2. Cheaper fuels (low octane fuel) can be used at high compression ratio.
3. Engine can use more than one fuel.
4. Excellent part load fuel economy.
5. Quite operation.
6. Load control can be achieved without air throttling.
7. Knock resistance smooth combustion.
8. Manufactured by existing technology.

Demerits

1. Maximum output is not achieved.
2. Speed ranges is inferior to petrol engine.
3. Expensive fuel injection system.
4. Complicated design.
5. Higher weight to power ratio.
6. Its reliability is yet to be established.

Applications

All these characteristics promise for the stratified charge engine a bright future because of its compact lightweight design, good fuel economy, speed range, antiknock resistance, multifuel capability and smooth and quiet engine operation over the entire speed and load range. A number of companies are actively working on this engine for automotive applications.

8.5   Gasoline Direct Injection (GDI)

In traditional gasoline engines, fuel is injected into a port before it enters the combustion chamber. The use of a port gives time to evaporate and mix uniformly with the air that it will be burnt with. However, this is not ideal because it provides little control over the air and fuel mixture. GDI engines use stratified charge, un-throttled operation to achieve improved part-load fuel economy over conventional Port fuel Injection (PFI) engines. In this case the injection of the fuel is done at the end of the compression phase.

These types of engines provide both the specific power of the gasoline engine and the efficiency of the diesel engine. One of the drawbacks of current GDI technology is that the fuel spray dynamics required for proper stratified charge operation usually results in liquid fuel impingement on combustion chamber surfaces.

The liquid fuel impingement is used to create a near-stoichiometic fuel/air mixture near the wall-mounted spark plug, but a fuel-lean mixture in the remaining volume of the combustion chamber. The process involved in GDI engine is shown in Fig.8.5.

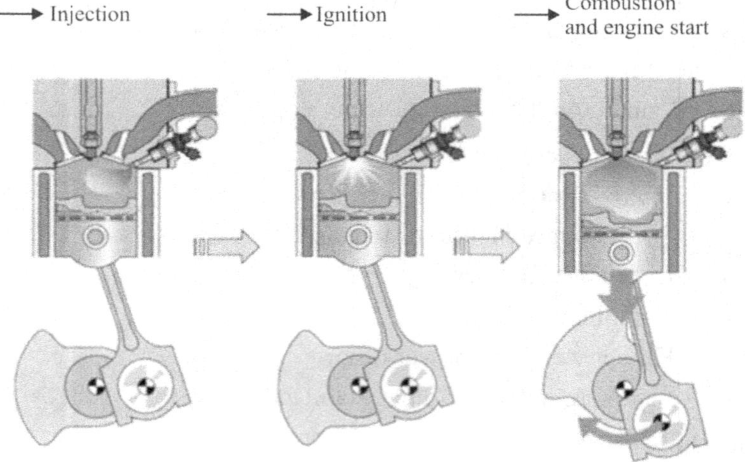

Fig. 8.5 GDI Engine

However a portion of the liquid fuel remains on the chamber surface and does not participate in the combustion process, resulting in high levels of unburned hydrocarbon on emissions and less-than-optimal fuel economy.

A laser ignition system can replace the wall-mounted spark plug and ignite the mixture at a location away from the chamber walls, negating the need for intentional fuel/wall interaction. This approach is expected to result in improved fuel economy and emissions.

Gasoline Direct Injection engines do not inject fuel into a port. These engines inject the fuel directly into the combustion chamber instead. This approach gives control over the air/fuel mixture and allows the engine to operate 2 modes as shown in Fig. 8.6.

Stratified Mode: Stratified mode is used when the engine is operating at low and medium loads. Fuel is injected into the chamber just before a spark ignites it. This allows very little time for the fuel to mix with air in the combustion chamber. As a result, the composition of the air and fuel mixture varies within the chamber. It is rich near the spark plug and lean in regions further away from it. A rich mixture is required near the spark plug to aid ignition. However, the overall composition of the mixture is lean and this reduces fuel consumption.

Homogeneous Mode: When a GDI engine needs to run at higher loads, it switches to a homogeneous mode to provide more power. For such situations, fuel is injected into the engine early during the induction stroke. This provides sufficient time for the fuel to mix uniformly with the air in the chamber and form a rich homogenous mixture for maximum power.

Major Objectives of the GDI engine

Ultra low fuel consumption or Superior power conventional MPI engines

1. *The differences between new GDI and current MPI:* For fuel supply, conventional engines use a fuel injecting system, which replace the fuel is injected to each intake port is currently the one of the most widely used systems. However, even in MPI engines there are limits to fuel supply response and the combustion control because the fuel mixes with air before entering the cylinder. Mitsubishi set out to push those limits by developing an engine where gasoline is directly injected into the cylinder as in a diesel engine, and moreover, where injection timings are precisely controlled to match load conditions.

   The GDI engine achieved the following outstanding characteristics.

   Extremely precise control of fuel supply to achieve fuel efficiency that exceeds that of diesel engine by enabling combustion of an ultra lean mixture supply.

   Very sufficient intake and relatively high compression ratio unique to the GDI engine deliver both performances and response that surpasses those of conventional MPI engines.

2. ***Technical features:*** Upright straight intake ports for optimal airflow control in the cylinder curved top pistons for better combustion.

   High pressure fuel pump to feed pressurized fuel into the injections.

   High Pressure swirl injectors for optimum air fuel mixture.

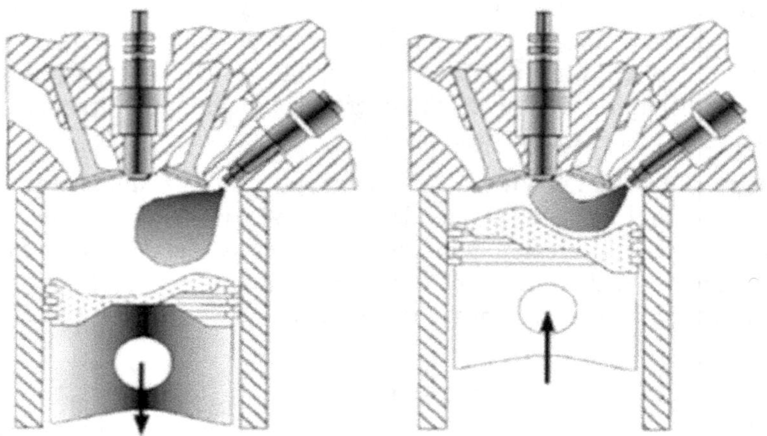

Homogeneous and stratified charge mode in a GDI engine

Fig. 8.6 GDI Engine in Homogenous and Stratified Mode

8.6 Overhead Cam Engine (OHC)

With the overhead cam engine, one or more camshafts located in the cylinder head open and close the valves for each cylinder.

Advantages of OHC

1. The OHC design is superior for the smaller, higher-revving engines found in passenger vehicles.
2. OHC allows a more efficient and compact design and more positive actuation of the valves.
3. OHC engines have fewer moving parts, so there is less inertia, or moving weight. Therefore the engines can be run at higher speeds and often produce more power than larger-capacity OHV engines.

SOHC

1. One camshaft operates both the intake and exhaust valves.
2. Camshaft is located in the cylinder head.

3. Valve train are arranged to operate directly through lifters or by rocker arms.

4. This configuration is very popular in modern automotive engines.

DOHC

1. Intake and exhaust valves each operate from separate camshafts directly through the lifters.

2. Provides more precise control for enhanced performance.

3. It is used in more expensive automotive applications.

4. Camshafts are present in V6 or V8 engines (2 for each cylinder head).

(a) SOHC Engine   **(b) DOHC Engine**

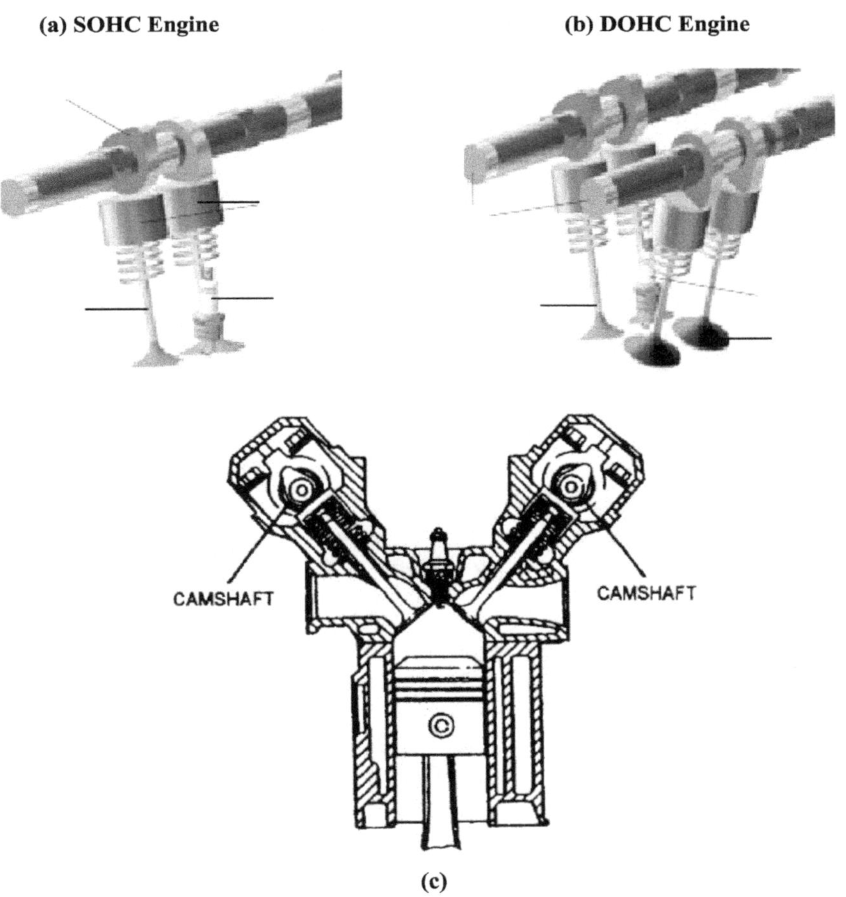

Fig. 8.7 Over Head Cam Engine

Review Questions

1. What is a lean burn engine? What are the advantages of using lean mixture in SI engine? What are the modifications to be made to convert an existing engine to operate on lean mixture?

2. What do you understand by Stratified Charge Engines? Discuss the different methods for stratification. Discuss the characteristic, advantages and disadvantages of Stratified charge IC engines.

3. Explain gasoline Direct Injection Engine.

4. With a neat sketch, explain plasma jet ignition system.

5. What do you mean by HCCI? Discuss the working principle of HCCI.

6. Write short notes on Overhead Cam Engine.

Index

C

About the Authors

K. Sudhakar obtained his B.E in Mechanical Engineering from Government College of Engineering, Salem, Madras University and M.Tech in Energy Management from School of Energy & Environment Studies, Devi Ahilya University, Indore, and Ph.D. from Centre for Energy and Environmental Science and Technology, CEESAT, National Institute of Technology, Tiruchirapalli, India. He is a certified Energy Manager & Auditor by BEE. He was awarded Senior Research fellowship from DST, Govt. of India for Doctoral Research and Young Scientist Award by Madhya Pradesh State Council of Science and Technology for his contribution in the field of Climate Change. He is Recipient of brightest Young Climate leader by British council, New Delhi. He has been actively involved in teaching & research in the area of Energy & Environment. His major research areas include Carbon Sequestration, Green Technology, Alternate fuels, Hydrogen fuel cell, nano and organic Solar cells, Solar Water Purification/desalination systems, Solar Thermal & PV, Energy modeling, Hybrid Energy systems, Exergy Analysis, Energy Conservation and management. He has published many research papers (around 70) both in referred national and international journals and conferences proceedings. He has been keynote speaker and resource person at several International and National conferences. He has visited several countries like Thailand, Srilanka, Indonesia for his research work. He has delivered several invited lectures in reputed institutes. He is currently working as Assistant Professor in Department of Energy at Maulana Azad National Institute of Technology (MANIT), Bhopal.

Dr. Anil Kumar presently works as Assistant Professor in Department of Energy, Maulana Azad National Institute of Technology (MANIT), Bhopal, India. Before joining MANIT, Bhopal, he was working as Lecturer in Department of Mechanical Engineering, University Institute of Technology, Rajiv Gandhi Proudyogiki Vishwavidyalaya, Bhopal (MP State Technological University). He has more than ten years of research and teaching experience in higher technical education.

He holds bachelor's degree in Mechanical Engineering followed by M.Tech. in Energy Technology and Ph.D. degree in solar thermal technologies from Indian Institute of Technology Delhi. Dr. Kumar's main area of interest is solar thermal technology, distribution of energy generation, clean energy technologies and application in buildings. He has published several books like, 'Energy Environment Ecology and Society', 'Fundamentals of Mechanical Engineering', 'Environmental Science: Fundamental, Ethics and Laws' and 'Advanced Internal Combustion Engines'. Dr. Kumar has published more than 46 research articles in journals and at International conferences and also 01 patent in his credit. Presently Dr. Kumar serves as reviewer to various international journals. In his tenure at MANIT, he has supervised ten M.Tech. students and is supervising six Ph.D. students. Dr. Kumar has carried various experimental projects in the area of solar water heating, solar drying for different crops, active and passive greenhouse dryer, solar collector assisted greenhouse dryers, and ground heat exchangers.